I0493834

Occurrence and Potential Transport of Selected Pharmaceuticals and Other Organic Wastewater Compounds from Wastewater-Treatment Plant Influent and Effluent to Groundwater and Canal Systems in Miami-Dade County, Florida

By Adam L. Foster, Brian G. Katz, and Michael T. Meyer

Prepared in cooperation with the Miami-Dade Water and Sewer Department and the Department of Environmental Resources Management

Series Name 2012–5083

U.S. Department of the Interior
U.S. Geological Survey

U.S. Department of the Interior
KEN SALAZAR, Secretary

U.S. Geological Survey
Marcia K. McNutt, Director

U.S. Geological Survey, Reston, Virginia: 2012

For more information on the USGS—the Federal source for science about the Earth, its natural and living resources, natural hazards, and the environment, visit http://www.usgs.gov or call 1–888–ASK–USGS.

For an overview of USGS information products, including maps, imagery, and publications,
visit http://www.usgs.gov/pubprod

To order this and other USGS information products, visit http://store.usgs.gov

Any use of trade, product, or firm names is for descriptive purposes only and does not imply endorsement by the U.S. Government.

Although this report is in the public domain, permission must be secured from the individual copyright owners to reproduce any copyrighted materials contained within this report.

Suggested citation:
Foster, A.L., Katz, B.G., and Meyer, M.T., 2012, Occurrence and potential transport of selected pharmaceuticals and other organic wastewater compounds from wastewater-treatment plant influent and effluent to groundwater and canal systems in Miami-Dade County, Florida: U.S. Geological Survey Scientific Investigations Report 2012–5083, 64 p, plus appendixes.

Contents

Figures

Tables

Conversion Factors and Abbreviations

Multiply	By	To obtain
Length		
inch (in.)	2.54	centimeter (cm)
inch (in.)	25.4	millimeter (mm)
foot (ft)	0.3048	meter (m)
mile (mi)	1.609	kilometer (km)
Volume		
gallon (gal)	3.785	liter (L)
gallon (gal)	0.003785	cubic meter (m^3)
million gallons (Mgal)	3,785	cubic meter (m^3)
Flow rate		
acre-foot per day (acre-ft/d)	0.01427	cubic meter per second (m^3/s)
acre-foot per year (acre-ft/yr)	1,233	cubic meter per year (m^3/yr)
acre-foot per year (acre-ft/yr)	0.001233	cubic hectometer per year (hm^3/yr)
foot per second (ft/s)	0.3048	meter per second (m/s)
cubic foot per second (ft^3/s)	0.02832	cubic meter per second (m^3/s)
cubic foot per second per square mile [(ft^3/s)/mi^2]	0.01093	cubic meter per second per square kilometer [(m^3/s)/km^2]
cubic foot per day (ft^3/d)	0.02832	cubic meter per day (m^3/d)
gallon per minute (gal/min)	0.06309	liter per second (L/s)
gallon per day (gal/d)	0.003785	cubic meter per day (m^3/d)
million gallons per day (Mgal/d)	0.04381	cubic meter per second (m^3/s)
Mass		
ounce, avoirdupois (oz)	28.35	gram (g)
pound, avoirdupois (lb)	0.4536	kilogram (kg)
Hydraulic conductivity		
foot per day (ft/d)	0.3048	meter per day (m/d)
Transmissivity*		
foot squared per day (ft^2/d)	0.09290	meter squared per day (m^2/d)

Temperature in degrees Celsius (°C) may be converted to degrees Fahrenheit (°F) as follows:

°F=(1.8×°C)+32

Concentrations of chemical constituents in water are given in micrograms per liter (µg/L).

Concentrations of chemical constituents in sediments are given in micrograms per kilogram (µg/kg).

Loads were calculated by multiplying the average flow during each 24-hour sampling period by total concentrations of constituents in the effluent and by a conversion factor (8.34×10^{-9}) to convert million gallons per day and micrograms per liter to pounds per day.

Chemical Abbreviations

3,4-DCA	3,4-dichloroaniline
AHTN	Tonalide
BHA	3-*tert*-butyl-4-hydroxyanisole
CIAT	2-Chloro-4-isopropylamino-6-amino-s-triazine
DCB	1,4-Dichlorobenzene
DEET	N-N-diethyl-*meta*-toluamide
DEHP	Diethylhexyl phthalate
DO	dissolved oxygen
E2	17-*beta*-Estradiol
HHCB	Galaxolide
NDMA	*N*-Nitrosodimethylamine
NP	4-Nonylphenol (total)
NP_1EO	4-Nonylphenol monoethoxylate
NP_2EO	4-Nonylphenol diethoxylate
OP	4-*tert*-Octylphenol
OP_1EO	4-Octylphenol monoethoxylate
OP_2EO	4-Octylphenol diethoxylate
SVOCs	Semivolatile organic compounds
TBEP	Tri(2-butoxyethyl) phosphate
TCEP	Tri(2-chloroethyl) phosphate
TDCP	Tri(dichloroisopropyl) phosphate
TBP	Tributylphosphate

Abbreviations

C	Unable to compute percent reduction value
CD1	Central District Wastewater Treatment Plant 1
CD2	Central District Wastewater Treatment Plant 2
CDWWTP	Central District Wastewater Treatment Plant
CLLE	Continuous liquid-liquid extraction
DERM	Miami-Dade Department of Environmental Resources Management
E	Estimated value
ELISA	Enzyme-linked immunosorbent assay
ESI	Electrospray ionization
GC	Gas chromatography
HPLC	High-performance liquid chromatography
hr	hour
HSWWTP	Homestead Wastewater Treatment Plant
lb/d	Pounds per day
LC	Liquid chromatography
MCL	Maximum contaminant level
MDL	Method detection level
Mgal/d	Million gallons per day
MRL	Method reporting level
MC	Mowry Canal
MS	Mass spectrometry
MW	monitoring well
µg/L	Microgram per liter
ND	North District Wastewater Treatment Plant – one plant only
NDWWTP	North District Wastewater Treatment Plant
NWQL	U.S. Geological Survey National Water Quality Laboratory
OGRL	U.S. Geological Survey Organic Geochemistry Research Laboratory
OLSPE	On-line solid-phase extraction
PCPs	Personal care products
QA	Quality assurance
RPD	Relative percent difference
SBR	Sequential batch reactor
SC	Snapper Creek
SD1	South District Wastewater Treatment Plant 1
SD2	South District Wastewater Treatment Plant 2
SDWWTP	South District Wastewater Treatment Plant
SPE	Solid-phase extraction
USEPA	U.S. Environmental Protection Agency
USGS	U.S. Geological Survey
WASD	Miami-Dade Water and Sewer Department
WWTP	Wastewater-treatment plant

Acknowledgments

The authors gratefully thank the many wastewater-treatment plant personnel for their assistance with sample collection. Special thanks are extended to Clive Powell, Steve Kronheim, Andy Toledo, Yvonne Walton, Manny Moncholi (Miami-Dade Water and Sewer Department), and Steve Anderson (City of Homestead) for their technical assistance at the facilities. Virginia Walsh, Sonia Villamil (Miami-Dade Water and Sewer Department) and Julie Baker (Miami-Dade Department of Environmental Resources Management) provided invaluable logistical assistance.

The authors benefited greatly from discussions with Pat Phillips, Steve Smith, and Dan Edwards (U.S. Geological Survey) on sampling and analysis of pharmaceuticals and other organic waste-water compounds. Carrie Boudreau, Amarys Acosta, and Michele Markovits (U.S. Geological Survey) assisted in sample collection and processing. Amy Gill and Steven Sando (U.S. Geological Survey) provided helpful reviews of this report.

Occurrence and Potential Transport of Selected Pharmaceuticals and Other Organic Wastewater Compounds from Wastewater-Treatment Plant Influent and Effluent to Groundwater and Canal Systems in Miami-Dade County, Florida

By Adam L. Foster, Brian G. Katz, and Michael T. Meyer

Abstract

An increased demand for fresh groundwater resources in South Florida has prompted Miami-Dade County to expand its water reclamation program and actively pursue reuse plans for aquifer recharge, irrigation, and wetland rehydration. The U.S. Geological Survey, in cooperation with the Miami-Dade Water and Sewer Department (WASD) and the Miami-Dade Department of Environmental Resources Management (DERM), initiated a study in 2008 to assess the presence of selected pharmaceuticals and other organic wastewater compounds in the influent and effluent at three regional wastewater-treatment plants (WWTPs) operated by the WASD and at one WWTP operated by the City of Homestead, Florida (HSWWTP).

To assess the range in concentrations of pharmaceuticals and other organic wastewater compounds in influent and effluent waters, 24-hour (hr) flow-weighted composite water samples were collected from the influent and effluent at each plant and analyzed for a broad range of pharmaceuticals and other organic wastewater compounds. These analyses included semivolatile organic compounds, pesticides and pesticide degradates, wastewater-indicator compounds, pharmaceuticals, antibiotics, and one hormone. Water samples were collected once in the wet season (high-flow conditions) and once in the dry season (low-flow conditions) to capture any seasonal variations in concentrations. Compounds detected in 24-hr flow-weighted influent composite samples included: 20 semivolatile organic compounds, 12 pesticides and pesticide degradates, 52 wastewater-indicator compounds, 5 pharmaceuticals, 14 antibiotics, and the hormone 17-*beta*-estradiol. Compounds detected in 24-hr flow-weighted effluent composite samples included: 19 semivolatile organic compounds, 13 pesticides, 49 wastewater-indicator compounds, 7 pharmaceuticals, 11 antibiotics, and 17-*beta*-estradiol. Percent reductions during treatment were calculated when possible. Among the various groups of compounds, semivolatile organic compounds and wastewater-indicator compounds generally had higher removal efficiencies than pharmaceuticals, antibiotics, and pesticides. Qualitatively, the total concentrations of pharmaceuticals and other organic wastewater compounds in the effluent at each plant were slightly higher in the dry season compared to the wet season; however, loads to the environment were generally larger in the wet season due to the higher flows through the WWTPs.

For decades, the HSWWTP has been discharging treated effluent directly to the water table using onsite soakage trenches. Water samples were collected from three monitoring wells at the HSWWTP to determine the occurrence of pharmaceuticals and other organic wastewater compounds in groundwater near the soakage trenches. Concentrations in the groundwater were generally below 0.5 microgram per liter (μg/L) with the exception of the fragrance galaxolide (estimated at 0.62–1.3 μg/L), the antibiotic sulfamethoxazole (0.14–0.57 μg/L), and the flame retardant tri(2-butoxyethyl) phosphate (estimated at 0.51 μg/L). Most pharmaceuticals and other organic wastewater compounds were attenuated in groundwater at this site; however, galaxolide, sulfamethoxazole, 3,4-dichloroaniline, 1,2-dichlorobenzene, and carbamazepine were detected in all groundwater samples, thus indicating that these compounds may be used as tracers of effluent at this site. Water samples were collected from monitoring wells once in the wet season and once in the dry season to determine any seasonal variations in concentrations. Results from seasonal sampling indicate that concentrations of pharmaceuticals and other organic wastewater compounds in groundwater are higher in the wet season, which is most likely related to the larger effluent loads to the groundwater during wet months.

Water and bed sediment samples were collected from two canal sites near the HSWWTP. These samples were analyzed for pharmaceuticals and other organic compounds to determine if compounds present in the effluent are being transported through the groundwater system and into the canal system. Water and bed sediment samples were also collected from one background canal site in Miami-Dade County, with no known sources of wastewater in the area, to compare with results from the canal near the HSWWTP. Results from these samples indicated that 51 compounds were detected in one or more canal samples with concentrations generally below 1.0 µg/L. A high estimated concentration of the plasticizer, diethylhexyl phthalate (DEHP, estimated at 11 µg/L), was detected in a canal water sample close to the HSWWTP; however, the source of this compound to the canal is most likely from surface runoff or the application of pesticides and not the HSWWTP. Results from seasonal sampling from the two canal sites indicated that concentrations of pharmaceuticals and other organic wastewater compounds were generally higher in the dry season. Water samples collected from the background canal site contained low levels (less than 0.76 µg/L) of 11 organic compounds in water samples, including 5 wastewater-indicator compounds and the non-prescription pharmaceutical acetaminophen; bed sediments contained detectable levels of 16 wastewater-indicator compounds. The presence of these compounds in water and bed sediment samples indicates that there is a probable nonpoint source of wastewater in the area.

Introduction

As demand for fresh groundwater resources increases, artificial recharge using highly treated wastewater, or reclaimed water, is becoming a widely utilized management technique to recharge aquifers in Florida and other parts of the United States (National Research Council, 1998; U.S. Environmental Protection Agency, 1994; Florida Department of Environmental Protection, 2010). An estimated 38 billion gallons of reclaimed water are used in Florida to recharge aquifers annually (WaterReuse Foundation, 2008). However, the use of reclaimed water for aquifer recharge in Miami-Dade County, a highly populated area on the southeastern coast of Florida, has been limited. Currently (2011), the City of Homestead operates the only wastewater-treatment plant (HSWWTP) in Miami-Dade County that uses reclaimed water for aquifer recharge (fig. 1). In 2008, an average of 5.17 million gallons per day (Mgal/d; or approximately 1.9 billion gallons total for the year) of tertiary-treated wastewater was used to recharge the Biscayne aquifer (Florida Department of Environmental Protection, 2010).

The Miami-Dade County Water and Sewer Department (WASD) is responsible for the collection, treatment, and disposal of the majority of the wastewater in the county and operates three regional wastewater-treatment plants (WWTPs) (North, Central, and South Districts, hereafter referred to as NDWWTP, CDWWTP, and SDWWTP, respectively) (fig. 1). These facilities dispose of approximately 300 Mgal/d (on average) of secondary-treated effluent through direct discharge of the treated effluent to ocean outfalls or through deep-well injection into the saline, confined Boulder Zone of the Lower Floridan aquifer. In 2008, the Florida legislature passed a bill setting a deadline for the elimination of existing wastewater ocean outfalls by 2025, with an added requirement that a majority of the wastewater previously discharged through this method be beneficially reused. In response to this bill, WASD is seeking to increase the amount of wastewater reused from its regional WWTPs, and is actively pursuing reuse plans to use reclaimed water for various purposes. For example, WASD is investigating the use of highly treated wastewater to restore historic freshwater flows to the Biscayne Bay coastal wetlands as part of the Comprehensive Everglades Restoration Plan. Also, a reclaimed water plant is under construction at the SDWWTP, which will recharge the Biscayne aquifer near the Zoo Miami with 21 Mgal/d of highly treated wastewater.

The Biscayne aquifer, a sole source aquifer comprising the uppermost part of the surficial aquifer system of southern Florida, is a highly permeable karst aquifer with estimates of transmissivity exceeding 300,000 square feet per day (ft^2/d; Fish and Stewart, 1991). Recent research has indicated the presence of stratiform, high-porosity zones of preferred flow within this carbonate system, which can serve as channels for potential rapid movement of contaminants (Cunningham and others, 2004; Renken and others, 2008). Embedded in the upper part of the surficial aquifer system is an interconnected system of man-made surface-water canals that serve a dual purpose. These canals (1) drain the Biscayne aquifer in the wet season to prevent flooding with discharge of excess water to Biscayne Bay and coastal wetlands (fig. 1), and (2) maintain heads in the aquifer in the dry season to prevent saltwater intrusion. Due to the high transmissivity of the Biscayne aquifer, the surface-water canals are directly connected to the groundwater system with aquifer responses being substantially influenced by canal operations.

Municipal wastewater contains many organic chemicals including prescription and non-prescription pharmaceuticals, human and veterinary antibiotics, pesticides, reproductive hormones, and other wastewater organic compounds. Although many are currently unregulated, some of these chemicals present known and potential health hazards to humans and other biota. In particular, pharmaceuticals, which are specifically designed to affect biological processes, have been identified as "Chemicals of Emerging Environmental Concern" by the U.S. Geological Survey (USGS) Toxic Substances Hydrology Program (U.S. Geological Survey, 2010). A critical concern when using reclaimed water, especially when reclaimed water is being used to recharge a public drinking-water supply, is to determine how effective treatment processes are in removing organic compounds from the water. Because the Biscayne aquifer is highly permeable, WASD and the Miami-Dade County Department of Environmental Resources Management (DERM) are concerned about the potential transport of these compounds from recharge areas to municipal supply well fields. Furthermore, because the Biscayne aquifer is interconnected with the canal system, there is potential for these compounds to be transported to the canal system and eventually

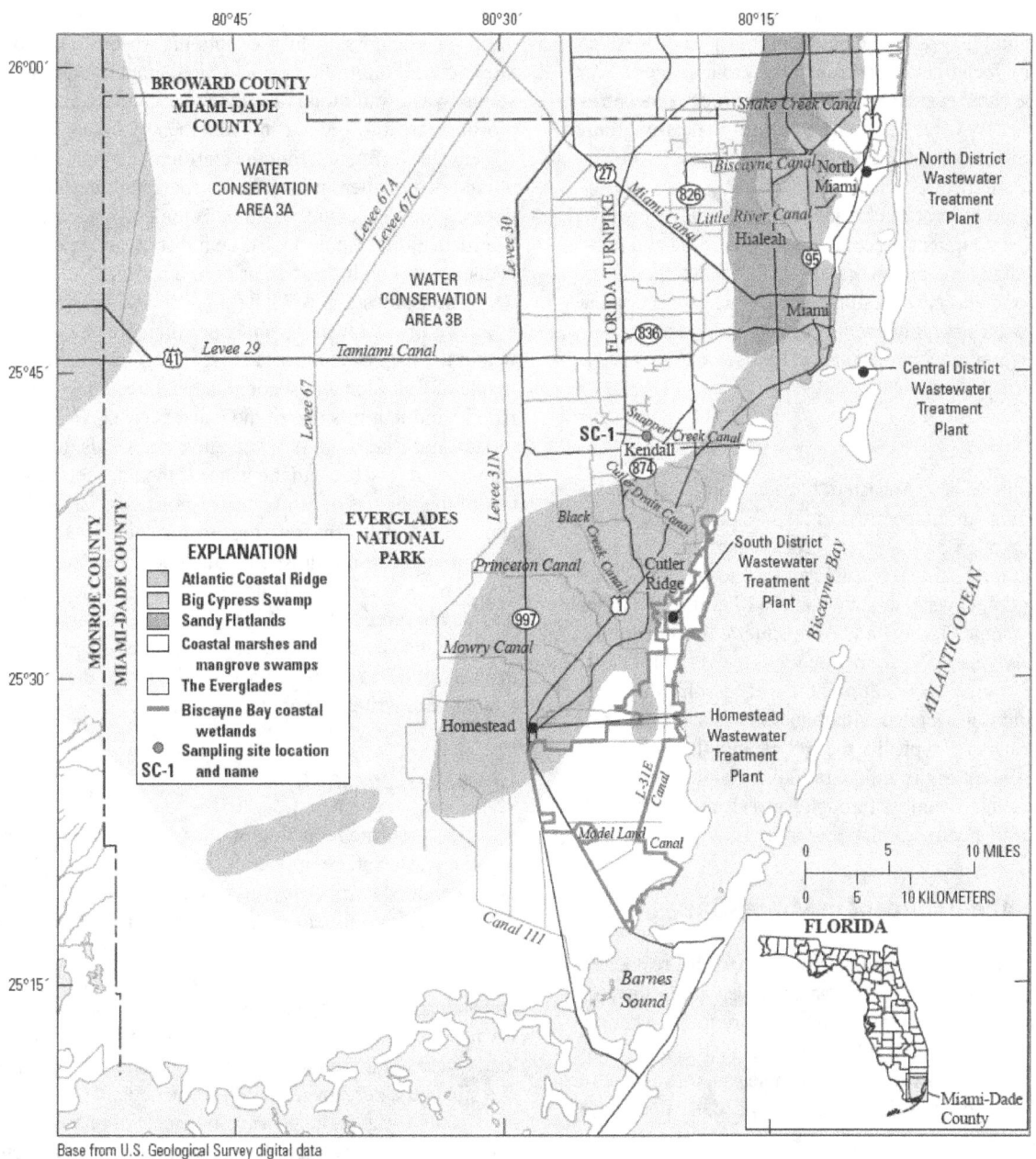

80°45'   80°30'   80°15'

26°00'

BROWARD COUNTY
MIAMI-DADE
COUNTY

WATER
CONSERVATION
AREA 3A

Snake Creek Canal

Biscayne Canal

North District
Wastewater
Treatment
Plant

WATER
CONSERVATION
AREA 3B

Little River Canal
Hialeah

Miami

25°45'

Levee 29   Tamiami Canal

Central District
Wastewater
Treatment
Plant

Snapper Creek Canal
SC-1
Kendall

EVERGLADES
NATIONAL
PARK

Black

Princeton Canal   Cutler
Ridge

South District
Wastewater
Treatment
Plant

Biscayne Bay

ATLANTIC OCEAN

EXPLANATION
Atlantic Coastal Ridge
Big Cypress Swamp
Sandy Flatlands
Coastal marshes and
  mangrove swamps
The Everglades
Biscayne Bay coastal
  wetlands
Sampling site location
SC-1   and name

25°30'

Mowry Canal

Homestead

Homestead
Wastewater
Treatment
Plant

Model Land Canal

0   5   10 MILES

0   5   10 KILOMETERS

Canal 111

FLORIDA

25°15'

Barnes
Sound

Miami-Dade
County

Base from U.S. Geological Survey digital data
Universal Transverse Mercator projection, Zone 17, Datum NAD 27

Figure 1. Miami-Dade County showing location of the North District Wastewater Treatment Plant, Central District Wastewater Treatment Plant, South District Wastewater Treatment Plant, Homestead Wastewater Treatment Plant, physiographic provinces, and the Biscayne Bay coastal wetlands in southeastern Florida.

to sensitive coastal estuarine and wetland systems (fig. 1). The USGS in cooperation with WASD and DERM initiated a study in 2008 to evaluate: (1) the presence of pharmaceuticals and other organic wastewater compounds in the influent and effluent at WASD's three regional WWTPs and at HSWWTP, (2) the percent reduction of compounds during treatment at each WWTP using current (2008) treatment processes, and (3) the occurrence of pharmaceuticals and other organic wastewater compounds in groundwater, canal water, and canal bed sediments near the HSWWTP.

Purpose and Scope

The purposes of this report are to: (1) assess and document the occurrence and concentrations of selected pharmaceuticals and other organic wastewater compounds in WWTP influent and effluent, groundwater, canal water, and canal sediment samples at selected sites in Miami-Dade County; (2) assess the persistence of these compounds during wastewater treatment by comparing influent and effluent concentrations, loads, and seasonal variations, and by calculating percent reduction of individual compounds during treatment at each plant; and (3) assess the potential transport of these compounds from recharge areas through the groundwater system and into the canal system by comparing effluent concentrations at the HSWWTP with concentrations in nearby groundwater, nearby canal water and bed sediments, and more distant canal water and bed sediments.

Background information about the WWTPs is provided, including the treatment processes used. Sampling and processing methods and the laboratory analytical methods are described, and analytical results are presented from dry-season and wet-season sampling events. Results of field and laboratory quality-assurance procedures are presented.

Climate

The climate in southeastern Florida is tropical with average annual air temperature of 24.7 °C and average annual precipitation of 58.53 inches (in.) for the period 1971–2000, measured at the National Oceanic and Atmospheric Administration (NOAA) weather station at the Miami International Airport (National Oceanic and Atmospheric Administration, 2004). An average of 43.05 in. (about 74 percent of the annual rainfall) of rainfall occurs during the wet season (May through October) and is associated with thunderstorms and tropical cyclones. August is typically the wettest month, with 8.63 in. (about 15 percent of the annual rainfall). Rainfall during the dry season (November through April) is relatively light, averaging only about 2.5 in. per month.

Wastewater Treatment in Miami-Dade County

The WWTPs in this study treat a mixture of residential, municipal, and industrial wastewater as well as stormwater runoff during rainfall events. Stormwater runoff can contain runoff from streets, parking lots, rooftops, parks, and lawns. The average daily flow during the wet season to each plant can be more than two times the annual average daily flow due to stormwater runoff.

The three Miami-Dade Water and Sewer Department facilities (NDWWTP, CDWWTP, and SDWWTP) vary in flow, population served, and ratio of residential to industrial wastewater treated. The NDWWTP consists of only one plant (ND) whereas the CDWWTP and SDWWTP consist of two separate plants that run in parallel. The two plants at the CDWWTP (CD1 and CD2) and SDWWTP (SD1 and SD2) were considered separate plants for this study because they receive wastewater from different areas and vary in flow (table 1). Combined, the three facilities are permitted to treat 368 Mgal/d of wastewater. The three facilities treat wastewater using an advanced treatment process known as "pure oxygen activated sludge." Briefly, raw wastewater (influent) enters the aerated grit chambers where particulate inorganic material settles and is collected for eventual transport to a disposal facility. Wastewater subsequently enters the aeration basins where 95-percent pure oxygen, generated by cryogenic oxygen generators, is diffused into the waste stream to facilitate bacterial multiplication and biodegradation of organic solids. Wastewater then flows to the secondary clarifiers where biosolids are allowed to settle and then removed. The treated wastewater (effluent) from the clarifiers then goes to the pump houses where it is either discharged to ocean outfalls or injected into the saline, confined Boulder Zone of the Lower Floridan aquifer (table 1). Effluent discharged to ocean outfalls is disinfected with chlorine prior to disposal.

The Homestead WWTP (HSWWTP) is permitted to treat 6 Mgal/d of wastewater from approximately 58,000 people within the City of Homestead. Treatment of wastewater begins in one of four identical sequential batch reactors (SBRs). Briefly, influent fills one of the reactors (tanks) and the wastewater is aerated. After aeration, the solids in the water are allowed to settle and the water is then decanted. While one of the tanks is in settle/decant mode, another is filling and aerating. Effluent from the tanks is then filtered using a sand/mixed media followed by disinfection of treated water by ultraviolet reactors. Treated effluent from the plant is then discharged directly to the Biscayne aquifer using rapid-rate adsorption soakage trenches (fig. 2). The average daily flow at the HSWWTP in 2009 was 5.1 Mgal/d (S. Anderson, City of Homestead, written commun., 2010).

General Approach

To assess the reduction of pharmaceuticals and other organic wastewater compounds during treatment, 24-hour (hr) flow-weighted composite water samples were collected from the influent and effluent at each plant (ND, CD1, CD2, SD1, SD2, and HS). These samples were collected on two occasions, once in the wet season and once in the dry season, and analyzed for selected pharmaceuticals and other organic wastewater compounds (table 2). Concentrations and total loads were compared for the influent and the effluent samples at a site and a percent reduction was calculated.

To investigate the possible transport of pharmaceuticals and other organic wastewater compounds in groundwater from the discharge of treated wastewater in the soakage trenches at the HSWWTP, three monitoring wells (MW) at varying distances from the soakage trenches (MW-1, MW-2, and MW-3; fig. 2) were sampled and analyzed for selected pharmaceuticals and other organic wastewater compounds (table 2). The monitoring wells were sampled once in the wet season and once in the dry season to determine if concentrations varied seasonally. Effluent and groundwater concentrations were compared as were wet-season and dry-season concentrations.

To investigate the possible transport of pharmaceuticals and other organic wastewater compounds through the groundwater system and into the canal system, canal water samples were collected from two sites near the HSWWTP in the Mowry Canal (MC-1, MC-2; fig. 2) and analyzed for selected pharmaceuticals and other organic wastewater compounds (table 2).

Table 1.   Summary of wastewater-treatment plants and sampling information for this study.

[Mgal/d, million gallons per day; WWTP, wastewater-treatment plant; SBR, Sequential Batch Reactor; POAS, pure oxygen acitvated sludge; F, filtration; UV, ultraviolet; Cl, chlorination; ST, soakage trenches; DWI, deep-well injection; OO, ocean outfall; —, none]

Facility name	Short name	Plant capacity (Mgal/d)	Secondary treatment	Tertiary treatment	Disinfection	Effluent disposal	Sample date (dry season)	Average flow during 24-hour sampling period (Mgal/d; dry season)	Sample date (wet season)	Average flow during 24-hour sampling period (Mgal/d; wet season)
Homestead WWTP	HS	6	SBR	F	UV	ST	2/10/2009	4.5	10/20/2009–10/21/2009	5.6
South District WWTP 1	SD1	112.5[1]	POAS	—	—	DWI	1/28/2009	53.2	10/13/2009–10/14/2009	62.1
South District WWTP 2	SD2	112.5[1]	POAS	—	—	DWI	1/28/2009	18.8	10/13/2009–10/14/2009	28.2
Central District WWTP 1	CD1	143[1]	POAS	—	Cl	OO	4/20/2009–4/21/2009	36.3	9/29/2009–9/30/2009	48.8
Central District WWTP 2	CD2	143[1]	POAS	—	Cl	OO	4/27/2009–4/28/2009	40.6	9/29/2009–9/30/2009	69.7
North District WWTP	ND	112.5	POAS	—	Cl	DWI, OO	12/9/2008–12/10/2008	47.0	9/22/2009–9/23/2009	90.4

[1]Combined flow

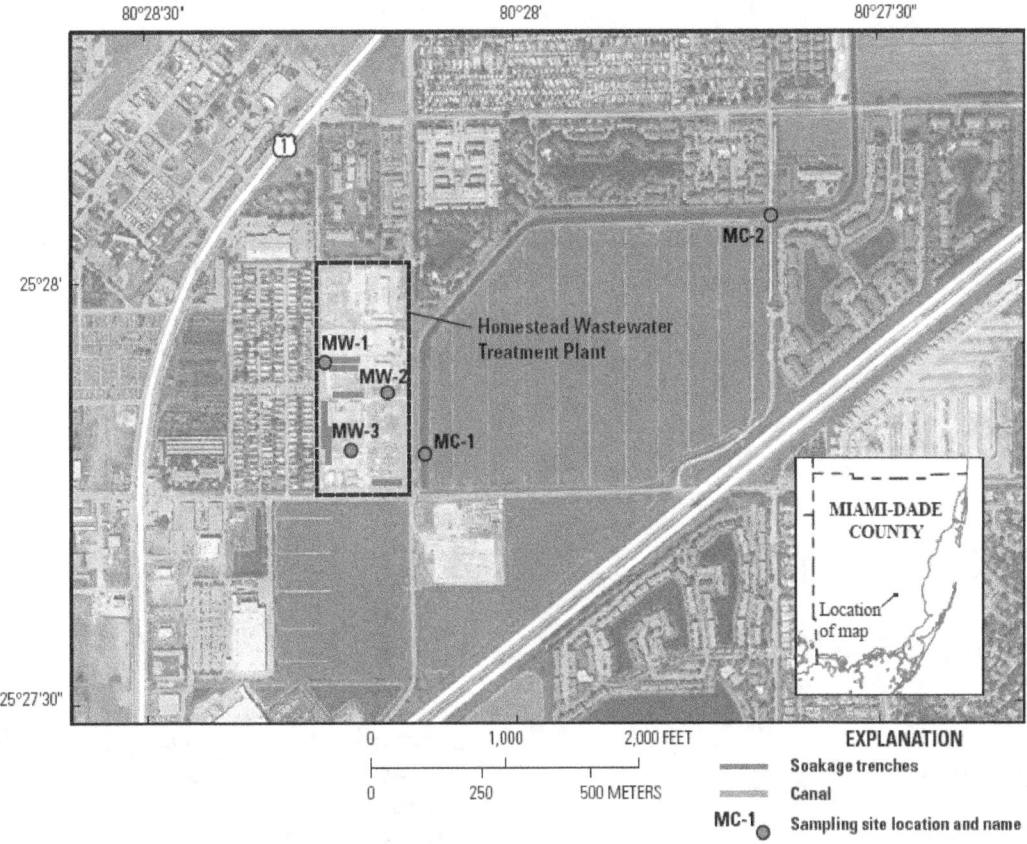

Figure 2.   Location of Homestead Wastewater Treatment Plant, monitoring wells and canal sampling site locations, and soakage trenches in the Miami-Dade County, Florida, study area.

Table 2. Pharmaceuticals and other organic wastewater compounds analyzed in water samples.

[CAS RN, Chemical Abstract Service Registry Number; MRL, method reporting level; μg/L, micrograms per liter; P-Code, U.S. Environmental Protection Agency STORET code; NWQL, U.S. Geological Survey National Water Quality Laboratory; CLLE, Continiuous liquid-liquid extraction; GC/MS, gas chromatogrpahy/mass spectrometry; OGRL, U.S. Geological Survey Organic Geochemistry Research Laboratory; SPE, solid-phase extraction; HPLC/MS, high-performance liquid chromatogrpahy/mass spectrometry; ELISA, enzyme-linked immunosorbent assay; —, not applicable]

Compound	CAS RN	MRL (μg/L)	P-Code
Semivolatile organic compounds (NWQL, Unfiltered, CLLE, GC/MS)			
1,2,4-Trichlorobenzene	120-82-1	0.26	34551
1,2-Dichlorobenzene	95-50-1	0.2	34536
1,2-Diphenylhydrazine	122-66-7	0.3	82626
1,3-Dichlorobenzene	541-73-1	0.22	34566
1,4-Dichlorobenzene	106-46-7	0.22	34571
2,4,6-Trichlorophenol	88-06-2	0.34	34621
2,4-Dichlorophenol	120-83-2	0.36	34601
2,4-Dimethylphenol	105-67-9	0.8	34606
2,4-Dinitrophenol	51-28-5	1	34616
2,4-Dinitrotoluene	121-14-2	0.56	34611
2,6-Dinitrotoluene	606-20-2	0.4	34626
2-Chloronaphthalene	91-58-7	0.16	34581
2-Chlorophenol	95-57-8	0.26	34586
2-Nitrophenol	88-75-5	0.4	34591
4-Bromophenylphenylether	101-55-3	0.24	34636
4-Chloro-3-methylphenol	59-50-7	0.54	34452
4-Chlorophenyl phenyl ether	7005-72-3	0.34	34641
4-Nitrophenol	100-02-7	0.52	34646
Acenaphthene	83-32-9	0.28	34205
Acenaphthylene	208-96-8	0.3	34200
Benz[a]anthracene	56-55-3	0.26	34526
Benzo[b]fluoranthene	205-99-2	0.3	34230
Benzo[ghi]perylene	191-24-2	0.38	34521
Benzo[k]fluoranthene	207-08-9	0.3	34242
Bis(2-chloroethoxy)methane	111-91-1	0.24	34278
Bis(2-chloroethyl)ether	111-44-4	0.3	34273
Bis(2-chloroisopropyl)ether	108-60-1	0.14	34283
Butylbenzyl phthalate	85-68-7	0.9	34292
Chrysene	218-01-9	0.32	34320
Dibenz[a,h]anthracene	53-70-3	0.42	34556
Diethyl phthalate	84-66-2	0.62	34336
Diethylhexyl phthalate (DEHP)	117-81-7	1.3	39100
Dimethyl phthalate	131-11-3	0.36	34341
Di-n-butyl phthalate	84-74-2	1	39110
Di-n-octyl phthalate	117-84-0	0.6	34596
Fluorene	86-73-7	0.34	34381
Hexachlorobenzene	118-74-1	0.3	39700
Hexachlorobutadiene	87-68-3	0.24	39702
Hexachlorocyclopentadiene	77-47-4	0.5	34386
Hexachloroethane	67-72-1	0.24	34396
Indeno[1,2,3-cd]pyrene	193-39-5	0.38	34403
Nitrobenzene	98-95-3	0.26	34447
N-Nitrosodimethylamine (NDMA)	62-75-9	0.24	34438
N-Nitrosodi-n-propylamine	621-64-7	0.4	34428
N-Nitrosodiphenylamine	86-30-6	0.28	34433
Pentachlorophenol	87-86-5	0.6	39032

Table 2. Pharmaceuticals and other organic wastewater compounds analyzed in water samples.—Continued

[CAS RN, Chemical Abstract Service Registry Number; MRL, method reporting level; µg/L, micrograms per liter; P-Code, U.S. Environmental Protection Agency STORET code; NWQL, U.S. Geological Survey National Water Quality Laboratory; CLLE, Continiuous liquid-liquid extraction; GC/MS, gas chromatogrpahy/mass spectrometry; OGRL, U.S. Geological Survey Organic Geochemistry Research Laboratory; SPE, solid-phase extraction; HPLC/MS, high-performance liquid chromatogrpahy/mass spectrometry; ELISA, enzyme-linked immunosorbent assay; —, not applicable]

Compound	CAS RN	MRL (µg/L)	P-Code
Pesticides and pesticide degradates (NWQL, Filtered, SPE, GC/MS)			
1-Naphthol	90-15-3	0.036	49295
2,6-Diethylaniline	579-66-8	0.003	82660
2-Chloro-2,6-diethylacetanilide	6967-29-9	0.01	61618
2-Chloro-4-isopropylamino-6-amino-s-triazine (CIAT)	6190-65-4	0.006	4040
2-Ethyl-6-methylaniline	24549-06-2	0.01	61620
3,4-Dichloroaniline	95-76-1	0.0042	61625
4-Chloro-2-methylphenol	1570-64-5	0.0046	61633
Acetochlor	34256-82-1	0.01	49260
Alachlor	15972-60-8	0.008	46342
Atrazine	1912-24-9	0.008	39632
Azinphos-methyl	86-50-0	0.12	82686
Azinphos-methyl-oxon	961-22-8	0.042	61635
Benfluralin	1861-40-1	0.014	82673
Carbaryl	63-25-2	0.06	82680
Chlorpyrifos	2921-88-2	0.0018	38933
Chlorpyrifos, oxygen analog	5598-15-2	0.03	61636
cis-Permethrin	61949-76-6	0.01	82687
Cyfluthrin	68359-37-5	0.016	61585
Cypermethrin	52315-07-8	0.02	61586
Dacthal	1861-32-1	0.0076	82682
Desulfinylfipronil	—	0.012	62170
Desulfinylfipronil amide	—	0.029	62169
Diazinon	333-41-5	0.006	39572
Diazinon, oxygen analog	962-58-3	0.005	61638
Dichlorvos	62-73-7	0.02	38775
Dicrotophos	141-66-2	0.08	38454
Dieldrin	60-57-1	0.008	39381
Dimethoate	60-51-5	0.006	82662
Ethion	563-12-2	0.008	82346
Ethion monoxon	17356-42-2	0.021	61644
Fenamiphos	22224-92-6	0.03	61591
Fenamiphos sulfone	31972-44-8	0.054	61645
Fenamiphos sulfoxide	31972-43-7	0.08	61646
Fipronil	120068-37-3	0.018	62166
Fipronil sulfide	120067-83-6	0.012	62167
Fipronil sulfone	120068-36-2	0.024	62168
Fonofos	944-22-9	0.0048	4095
Hexazinone	51235-04-2	0.008	4025
Iprodione	36734-19-7	0.014	61593
Isofenphos	25311-71-1	0.006	61594
Malaoxon	1634-78-2	0.022	61652
Malathion	121-75-5	0.016	39532
Metalaxyl	57837-19-1	0.014	61596
Methidathion	950-37-8	0.012	61598
Metolachlor	51218-45-2	0.02	39415
Metribuzin	21087-64-9	0.012	82630

Table 2. Pharmaceuticals and other organic wastewater compounds analyzed in water samples.—Continued

[CAS RN, Chemical Abstract Service Registry Number; MRL, method reporting level; µg/L, micrograms per liter; P-Code, U.S. Environmental Protection Agency STORET code; NWQL, U.S. Geological Survey National Water Quality Laboratory; CLLE, Continiuous liquid-liquid extraction; GC/MS, gas chromatogrpahy/mass spectrometry; OGRL, U.S. Geological Survey Organic Geochemistry Research Laboratory; SPE, solid-phase extraction; HPLC/MS, high-performance liquid chromatogrpahy/mass spectrometry; ELISA, enzyme-linked immunosorbent assay; —, not applicable]

Compound	CAS RN	MRL (µg/L)	P-Code
Pesticides and pesticide degradates (NWQL, Filtered, SPE, GC/MS)			
Myclobutanil	88671-89-0	0.01	61599
Paraoxon-methyl	950-35-6	0.014	61664
Parathion-methyl	298-00-0	0.008	82667
Pendimethalin	40487-42-1	0.012	82683
Phorate	298-02-2	0.02	82664
Phorate oxygen analog	2600-69-3	0.027	61666
Phosmet	732-11-6	0.07	61601
Phosmet oxon	3735-33-9	0.0079	61668
Prometon	1610-18-0	0.012	4037
Prometryn	7287-19-6	0.006	4036
Propyzamide	23950-58-5	0.0036	82676
Simazine	122-34-9	0.006	4035
Tebuthiuron	34014-18-1	0.028	82670
Terbufos	13071-79-9	0.018	82675
Terbufos oxygen analog sulfone	56070-15-6	0.045	61674
Terbuthylazine	5915-41-3	0.006	4022
Tribufos	78-48-8	0.018	61610
Trifluralin	1582-09-8	0.018	82661
Wastewater-indicator compounds (NWQL, Unfiltered, CLLE, GC/MS)			
1-Methylnaphthalene	90-12-0	0.02	81696
2,2',4,4'-Tetrabromodiphenylether (PBDE 47)	5436-43-1	0.02	63147
2,6-Dimethylnaphthalene	581-42-0	0.02	62805
2-Methylnaphthalene	91-57-6	0.02	30194
3,4-Dichlorophenyl isocyanate	102-36-3	0.16	63145
3-*beta*-Coprostanol	360-68-9	0.38	62806
3-Methyl-1H-indole (Skatol)	83-34-1	0.02	62807
3-*tert*-Butyl-4-hydroxyanisole (BHA)	25013-16-5	0.08	61702
4-Cumylphenol	599-64-4	0.02	62808
4-n-Octylphenol	1806-26-4	0.01	62809
4-Nonylphenol (total, NP)	84852-15-3	1.2	62829
4-Nonylphenol diethoxylate (NP$_2$EO)	26027-38-2	0.8	61703
4-Nonylphenol monoethoxylate (NP$_1$EO)	104-35-8	1.3	61704
4-Octylphenol diethoxylate (OP$_2$EO)	26636-32-8	0.1	62486
4-Octylphenol monoethoxylate (OP$_1$EO)	26636-32-8	0.3	62485
4-*tert*-Octylphenol (OP)	140-66-9	0.11	62810
5-Methyl-1H-benzotriazole	136-85-6	0.16	61944
Acetophenone	98-86-2	0.07	62811
Anthracene	120-12-7	0.01	34220
Anthraquinone	84-65-1	0.02	62813
Benzo[a]pyrene	50-32-8	0.01	34247
Benzophenone	119-61-9	0.08	62814
beta-Sitosterol	83-46-5	0.8	62815
beta-Stigmastanol	19466-47-8	0.8	61948
Bisphenol A	80-05-7	0.02	62816
Bromacil	314-40-9	0.08	30234

Table 2. Pharmaceuticals and other organic wastewater compounds analyzed in water samples.—Continued

[CAS RN, Chemical Abstract Service Registry Number; MRL, method reporting level; µg/L, micrograms per liter; P-Code, U.S. Environmental Protection Agency STORET code; NWQL, U.S. Geological Survey National Water Quality Laboratory; CLLE, Continiuous liquid-liquid extraction; GC/MS, gas chromatogrpahy/mass spectrometry; OGRL, U.S. Geological Survey Organic Geochemistry Research Laboratory; SPE, solid-phase extraction; HPLC/MS, high-performance liquid chromatogrpahy/mass spectrometry; ELISA, enzyme-linked immunosorbent assay; —, not applicable]

Compound	CAS RN	MRL (µg/L)	P-Code
Wastewater-indicator compounds (NWQL, Unfiltered, CLLE, GC/MS)			
Bromoform	75-25-2	0.08	32104
Camphor	76-22-2	0.04	62817
Carbazole	86-74-8	0.01	77571
Cholesterol	57-88-5	0.3	62818
Cotinine	486-56-6	0.04	61945
Fluoranthene	206-44-0	0.01	34376
Galaxolide (HHCB)	1222-05-5	0.04	62823
Indole	120-72-9	0.02	62824
Isoborneol	124-76-5	0.04	62825
Isophorone	78-59-1	0.02	34408
Isopropylbenzene	98-82-8	0.02	77223
Isoquinoline	119-65-3	0.02	62826
Limonene	5989-27-5	0.2	34336
Menthol	89-78-1	0.16	62827
Methyl salicylate	119-36-8	0.04	62828
N,N-diethyl-meta-toluamide (DEET)	134-62-3	0.02	61947
Naphthalene	91-20-3	0.01	34696
para-Cresol	106-44-5	0.04	77146
Phenanthrene	85-01-8	0.01	34461
Phenol	108-95-2	0.08	34694
Pyrene	129-00-0	0.01	34469
Tetrachloroethylene	127-18-4	0.08	34475
Tonalide (AHTN)	21145-77-7	0.02	62812
Tri(2-butoxyethyl) phosphate (TBEP)	78-51-3	0.32	62830
Tri(2-chloroethyl) phosphate (TCEP)	115-96-8	0.08	62831
Tri(dichlorisopropyl) phosphate (TDIP)	13674-87-8	0.16	61707
Tributyl phosphate (TPB)	126-73-8	0.02	62832
Triclosan	3380-34-5	0.16	61708
Triethyl citrate (Ethyl citrate)	77-93-0	0.02	62833
Triphenyl phosphate	115-86-6	0.04	62834
Pharmaceutical compounds (NWQL, Filtered, SPE, HPLC/MS)			
1,7-Dimethylxanthine (Wastewater-indicator compound)	611-59-6	0.1	62030
Acetaminophen	103-90-2	0.12	62000
Albuterol	18559-94-9	0.08	62020
Caffeine (Wastewater-indicator compound)	58-08-2	0.06	50305
Codeine	76-57-3	0.046	62003
Dehydronifedipine	67035-22-7	0.08	62004
Diltiazem	42399-41-7	0.02	62008
Diphenhydramine	147-24-0	0.058	62796
Thiabendazole	148-79-8	0.06	62801
Warfarin	81-81-2	0.08	62024

Table 2. Pharmaceuticals and other organic wastewater compounds analyzed in water samples.—Continued

[CAS RN, Chemical Abstract Service Registry Number; MRL, method reporting level; μg/L, micrograms per liter; P-Code, U.S. Environmental Protection Agency STORET code; NWQL, U.S. Geological Survey National Water Quality Laboratory; CLLE, Continiuous liquid-liquid extraction; GC/MS, gas chromatogrpahy/mass spectrometry; OGRL, U.S. Geological Survey Organic Geochemistry Research Laboratory; SPE, solid-phase extraction; HPLC/MS, high-performance liquid chromatogrpahy/mass spectrometry; ELISA, enzyme-linked immunosorbent assay; —, not applicable]

Compound	CAS RN	MRL (µg/L)	P-Code
Antibiotics (OGRL, Filtered, SPE, LC/MS)			
Fluoroquinolines			
Ciprofloxacin	85721-33-1	0.005	—
Enrofloxacin	93106-60-6	0.005	—
Lomefloxacin	98079-51-7	0.005	—
Norfloxacin	70458-96-7	0.005	—
Ofloxacin	82419-36-1	0.005	—
Sarafloxacin	98105-99-8	0.005	—
Macrolides and degradation products			
Azithromycin	83905-01-5	0.005	—
Erythromycin	114-07-8	0.008	—
Erythromycin-H2O	643-22-1	0.008	—
Roxithromycin	80214-83-1	0.005	—
Tylosin	1401-69-0	0.01	—
Virginiamycin	11006-76-1	0.005	—
Sulfonamides			
Sulfachloropyridazine	80-32-0	0.005	—
Sulfadiazine	68-35-9	0.1	—
Sulfadimethoxine	122-11-2	0.005	—
Sulfamethoxazole	723-46-6	0.005	—
Sulfamethazine	57-68-1	0.005	—
Suflathiazole	72-14-0	0.05	—
Tetracyclines and degradation products			
Chlortetracyline	57-62-5	0.01	—
Oxytetracycline	79-57-2	0.01	—
Tetracycline	60-54-8	0.01	—
Doxycycline	564-25-0	0.01	—
epi-Chlortetracycline	—	0.01	—
epi-iso-Chlortetracycline	—	0.01	—
epi-Oxytetracycline	—	0.01	—
epi-Tetracycline	—	0.01	—
iso-Chlortetracycline	—	0.01	—
Others			
Carbamazepine (Pharmaceutical)	298-46-4	0.005	—
Chloramphenicol	56-75-7	0.1	—
Ibuprofen (Pharmaceutical)	15687-27-1	0.005	—
Lincomycin	154-21-2	0.005	—
Ormetoprim	6981-18-6	0.005	—
Trimethoprim	738-70-5	0.005	—
Hormone (OGRL, Filtered, ELISA)			
17-*beta*-Estradiol (E2)	50-28-2	0.0015	—

MC-1 is located directly adjacent to the HSWWTP, whereas MC-2 is approximately 1 mile (mi) downstream of MC-1. Canal water samples were also collected from a location in the Snapper Creek Canal (SC-1; fig. 1). This site was used as a background site because there are no known sources of wastewater to this canal. MC-1 and MC-2 were sampled in both the wet and dry season, while SC-1 was only sampled in the wet season. Bed sediment samples were collected at the same locations and times as canal water samples (MC-1, MC-2, and SC-1), and analyzed for the wastewater-indicator compounds listed in table 3.

Previous Studies

Pharmaceuticals and other organic wastewater compounds are present in treated wastewater effluent and have been detected in surface water and groundwater near known sources of these contaminants in the United States and other parts of the world. Sando and others (2005) measured 50–60 pharmaceuticals and other organic wastewater compounds in treated wastewater effluent in South Dakota. Lietz and Meyer (2006) reported that 30–35 pharmaceuticals and other organic wastewater compounds were present in treated effluent from Miami-Dade County, Florida. These compounds also have been detected in surface waters that receive wastewater discharges and some of these compounds could be sufficiently mobile to be transported to the groundwater. Kolpin and others (2002) detected about 80 pharmaceuticals and other wastewater organic compounds in water from streams in the United States near known contaminant sources. Cordy and others (2004) determined in a laboratory study that some pharmaceuticals and other organic wastewater compounds, including carbamazepine, sulfamethoxazole, benzophenone, 5-methyl-1H-benzotriazole, N-N-diethyl-*meta*-toluamide (DEET), tributylphosphate, and tri(2-chloroethyl) phosphate (TCEP), may potentially reach the groundwater. Studies by Heberer and others (2004) and Barnes and others (2004) indicated that some organic compounds, such as carbamazepine, DEET, and TCEP, can be transported within groundwater systems. Heberer and others (2004) detected in public-supply wells carbamazepine and several other pharmaceuticals, which were transported from a nearby stream by bank filtration. Barnes and others (2004) detected about 20 pharmaceuticals and other organic wastewater compounds, including DEET, in wells downgradient from a municipal landfill.

Pharmaceuticals include antibiotics, hormones, analgesics, opioids, barbiturates, stimulants and other chemicals that constitute a special class of organic compounds because they are specifically synthesized for human or veterinary use. As part of a larger group of "emerging contaminants," the USGS Toxic Substances Hydrology Program (*http://toxics.usgs.gov/index.html and http://toxics.usgs.gov/regional/emc/index.html*) is engaged in a variety of research activities documenting the sources and persistence of pharmaceuticals in hydrologic systems (Barnes and others, 2008; Kinney and others, 2006;

Phillips and others, 2010), developing analytical techniques to detect and quantify these compounds (Cahill and others, 2004), and assessing the unintended effects of these compounds on organisms exposed to these compounds (Barber and others, 2007; Schultz and others, 2010; Vajda and others, 2008; Writer and others, 2001).

Methods

The following sections describe the methods used to collect and analyze data for this report. Methods described include sample collection and processing, analytical methods, quality assurance, and estimation of loads.

Sample Collection and Processing

All samples for this study were collected and processed according to standard USGS protocols (U.S. Geological Survey, variously dated). The 24-hr flow-weighted composite samples (2-hr intervals) were collected from the influent and effluent at the six plants (ND, CD1, CD2, SD1, SD2, and HSWWTP) using a Lagrangian sampling scheme (the initiation of sample collection at the effluent was lagged by hydraulic retention time, usually 2 to 4 hrs). At the Miami-Dade WWTPs, facility personnel collected the samples in precleaned 1-liter (L) glass bottles and immediately stored them in a refrigerator at 4 °C. At the HSWWTP, an ISCO (6712) automated sampler outfitted with fluoropolymer (Teflon®) tubing was used to collect influent and effluent samples. Ice blocks were placed in the samplers to keep the samples chilled before being processed. At the end of each 24-hr sample period, water samples were transported on ice to the water-quality laboratory at the USGS Water Science Center in Fort Lauderdale, Fla., for processing. Samples for each respective site and type were then composited in either a 14-L Teflon® churn splitter or a 10 gallon (gal) stainless steel bucket. For those samples requiring filtration, water was passed through a 0.7 micrometer, baked (450 °C for 8 hrs), glass-fiber filter.

All equipment was decontaminated and the laboratory area was cleaned and lined with aluminum foil prior to processing, following previous methods (Lietz and Meyer, 2006). The initial step in the decontamination procedure involved cleaning the equipment with a dilute phosphate-free detergent. This was followed by rinsing with: (1) deionized water, (2) pesticide-grade methanol, and (3) pesticide-grade organic-free water. The equipment was then wrapped in aluminum foil and sealed in plastic bags prior to use. Disposable powderless nitrile gloves, a laboratory coat, and safety glasses were worn during the processing and decontamination procedure.

Groundwater samples from the three monitoring wells at the HSWWTP (fig. 2) were collected using a portable submersible pump. The pump was constructed of stainless steel and outfitted with Teflon® tubing to minimize cross contamination from one well site to another. A minimum of

Table 3. Wastewater-indicator compounds analyzed in bed sediment samples.

[CAS RN, Chemical Abstract Service Registry Number; MRL, method reporting level; µg/kg, micrograms per kilogram; P-Code, U.S. Environmental Protection Agency STORET code]

Compound	CAS RN	MRL (µg/L)	P-Code	Use/Source
1,4-Dichlorobenzene	106-46-7	27	63163	Moth repellent, fumigant, deodorant
1-Methylnaphthalene	90-12-0	27	63165	PAH
2,2',4,4'-Tetrabromodiphenylether (PBDE 47)	5436-43-1	19	63166	Fire retardant
2,6-Dimethylnaphthalene	581-42-0	24	63167	Alkyl-PAH
2-Methylnaphthalene	91-57-6	27	63168	PAH
3-*beta*-Coprostanol	360-68-9	36	63170	Biogenic sterol
3-Methyl-1H-indole (Skatol)	83-34-1	30	63171	Fecal indicator
3-*tert*-Butyl-4-hydroxyanisole (BHA)	25013-16-5	10	63172	Antioxidant
4-Cumylphenol	599-64-4	33	63173	Nonionic detergent metabolite
4-*n*-Octylphenol	1806-26-4	36	63174	Nonionic detergent metabolite
4-Nonylphenol (total, NP))	104-40-5	49	63175	Nonionic detergent metabolite
4-Nonylphenol diethoxylate (NP$_2$EO)	26027-38-2	85	63200	Nonionic detergent metabolite
4-Nonylphenol monoethoxylate (NP$_1$EO)	104-35-8	33	63221	Nonionic detergent metabolite
4-Octylphenol diethoxylate (OP$_2$EO)	26636-32-8	38	63201	Nonionic detergent metabolite
4-Octylphenol monoethoxylate (OP$_1$EO)	26636-32-8	21	63206	Nonionic detergent metabolite
4-*tert*-Octylphenol (OP)	140-66-9	22	63176	Nonionic detergent metabolite
Acetophenone	98-86-2	10	63178	Fragrance
Anthracene	120-12-7	19	63180	PAH
Anthraquinone	84-65-1	24	63181	Manufacturing of dyes/textiles
Atrazine	1912-24-9	58	63182	Pesticide
Benzo[a]pyrene	50-32-8	24	63183	PAH
Benzophenone	119-61-9	31	63184	Fragrance
beta-Sitosterol	83-46-5	36	63185	Biogenic sterol
beta-Stigmastanol	19466-47-8	36	63186	Biogenic sterol
Bisphenol A	80-05-7	31	63188	Plasticizer
Bromacil	314-40-9	25	63189	Pesticide
Camphor	76-22-2	27	63192	Flavorant
Carbazole	86-74-8	22	63194	Pesticide
Chlorpyrifos	2921-88-2	33	63195	Pesticide
Cholesterol	57-88-5	16	63196	Biogenic sterol
Diazinon	333-41-5	48	63198	Pesticide
Diethyl phthalate	84-66-2	46	63202	Plasticizer
Diethylhexyl phthalate (DEHP)	117-81-7	13	63187	Plasticizer
Fluoranthene	206-44-0	23	63208	PAH
Galaxolide (HHCB)	1222-05-5	16	63209	Fragrance
Indole	120-72-9	53	63210	Fragrance
Isoborneol	124-76-5	39	63211	Fragrance
Isophorone	78-59-1	43	63212	Industrial solvent
Isopropylbenzene (Cumene)	98-82-8	86	63213	Solvent
Isoquinoline	119-65-3	83	63214	Flavorant
Limonene	5989-27-5	23	63203	Fragrance
Menthol	89-78-1	42	63215	Flavorant
Metolachlor	51218-45-2	37	63218	Pesticide
N,N-Diethyl-*meta*-toluamide (DEET)	134-62-3	56	63219	Pesticide
Naphthalene	91-20-3	23	63220	PAH
para-Cresol	106-44-5	16	63222	Wood preservative
Phenanthrene	85-01-8	20	63224	PAH
Phenol	108-95-2	38	63225	Disinfectant

Table 3.   Wastewater-indicator compounds analyzed in bed sediment samples.—Continued

[CAS RN, Chemical Abstract Service Registry Number; MRL, method reporting level; µg/kg, micrograms per kilogram; P-Code, U.S. Environmental Protection Agency STORET code]

Compound	CAS RN	MRL (µg/L)	P-Code	Use/Source
Prometon	1610-18-0	44	63226	Pesticide
Pyrene	129-00-0	20	63227	PAH
Tonalide (AHTN)	21145-77-7	12	63179	Fragrance
Tri(2-butoxyethyl) phosphate (TBEP)	78-51-3	98	63229	Fire retardant
Tri(2-chloroethyl) phosphate (TCEP)	115-96-8	70	63230	Fire retardant
Tri(dichloroisopropyl) phosphate (TDIP)	13674-87-8	73	63235	Fire retardant
Tributyl phosphate (TBP)	126-73-8	39	63231	Fire retardant
Triclosan	3380-34-5	49	63232	Disinfectant
Triphenyl phosphate	115-86-6	46	63234	Plasticizer

three casing volumes was purged from each well, and water samples were collected after field properties (temperature, pH, specific conductance, and dissolved oxygen) had stabilized. To minimize contamination from the atmosphere, sample chambers were used while collecting and processing ground-water samples. Sampling equipment was cleaned immediately after sample collection using the decontamination procedure described in the previous paragraph.

Samples of canal water were collected using a depth-integrating Teflon® sampler from 5-10 verticals and composited in a 14-L Teflon® churn splitter. Samples of bed sediments were collected using a stainless steel Ekman grab sampler (model #316ss). Each bed sediment sample consisted of approximately 35 cubic inches of sediment from the top 4 in. of the soil column. To minimize cross contamination from one canal site to another, both samplers were decontaminated immediately before and after the collection of each sample using the decontamination procedure described previously.

Samples collected for analysis of low levels (parts per trillion) of pharmaceuticals and other organic wastewater compounds are susceptible to contamination because these compounds are found in many products that are routinely used by field personnel. To ensure sample integrity during sample collection and preparation, field personnel avoided contact with or consumption of products that contain the compounds targeted for analysis. These products included soaps and detergents, insect repellents, fragrances, sunscreen, caffeine, and tobacco products. Field personnel also avoided contact with the following pharmaceuticals: prescription drugs, medications and hormonal substances, nonprescription medications, and selected human and veterinary antibiotics. Powderless nitrile gloves were frequently changed during sampling activities and bottling to prevent contamination. Direct contact between samples and clothing also was avoided during sampling and processing activities.

Analytical Methods

The concentrations of 210 organic compounds were determined using six different analytical methods. The 210 compounds included semivolatile organic compounds (SVOCs), pesticides and pesticide degradates, wastewater-indicator compounds, pharmaceuticals, antibiotics, and one hormone (17-*beta*-estradiol, E2) (table 2). The analytical procedures used for the determination of SVOCs, pesticides and pesticide degradates, wastewater-indicator compounds, and pharmaceuticals in water samples are approved USGS methods (Fishman, 1993; Zaugg and others, 1995; Zaugg and others, 2006; and Furlong and others, 2008). Antibiotics were analyzed using a method modified from Meyer and others (2007). The procedure used for the determination of E2 in water samples is a provisional research method not yet approved by the USGS.

The SVOCs, pesticides and pesticide degradates, wastewater-indicator compounds, and pharmaceuticals were analyzed at the USGS National Water Quality Laboratory (NWQL) in Denver, Colo. SVOCs were extracted from unfiltered water samples using continuous liquid-liquid extraction (CLLE). The extracts were analyzed for 46 SVOCs (table 2) and 5 surrogate compounds by gas chromatography/mass spectrometry (GC/MS) methods (Fishman 1993). Pesticides and pesticide degradates were extracted from filtered water samples using C-18 solid-phase extraction (SPE) cartridges. The extracts were analyzed for 64 pesticides and pesticide degradates (table 2) and 2 surrogate compounds by capillary-column GC/MS methods (Zaugg and others, 1995). Wastewater-indicator compounds were extracted from unfiltered water samples using CLLE. The extracts were analyzed for 56 wastewater-indicator compounds (table 2) and 4 surrogate compounds by GC/MS methods (Zaugg and others, 2006). Pharmaceuticals were extracted from filtered water samples using chemically modified styrene-divinylbenzene resin-based SPE cartridges. The extracts were analyzed for 8 pharmaceuticals, 2 wastewater-indicator compounds (caffeine and the caffeine metabolite, 1,7-dimethylxanthine) (table 2), and 2 surrogate compounds by high-performance liquid chromatography/mass spectrometry (HPLC/MS) methods (Furlong and others, 2008). Wastewater-indicator compounds were extracted from sediment samples using pressurized solvent extraction followed by SPE. The extracts were analyzed for 57 wastewater-indicator compounds (table 3) and 3 surrogate compounds by GC/MS methods (Burkhardt and others, 2006).

A total of 31 antibiotics, 2 pharmaceuticals (carbamazepine and ibuprofen), 3 surrogate compounds, and E2 were analyzed at the USGS Organic Geochemistry Laboratory (OGRL) in Lawrence, Kansas. The antibiotics and pharmaceuticals were analyzed using on-line solid-phase extraction (OLSPE) and liquid chromatography/tandem mass spectrometry (LC/MS/MS) modified from the OLSPE LC/MS method of Meyer and others, 2007. The analysis of E2 in filtered water samples was performed using commercially available magnetic-particle enzyme-linked immunosorbent assay (ELISA) kits (Abraxis, Warminster, Pennsylvania). Generally, the results from the ELISA method in complex matrices such as the ones analyzed in this study should be considered semi-quantitative (Farre and others, 2006) and, therefore, interpreted with some caution.

Many of the concentrations reported herein were below analytical method reporting levels (generally less than 0.4 µg/L); however, because they met qualitative criteria used for compound identification (chromatographic retention times, full-scan mass spectra, and ion abundance ratios), they were reported by the laboratory as estimated (E) values. Other factors that could lead to an estimated value include: (1) the sample matrix interfered with measurement of the compound; (2) surrogate recoveries (described in the next section) indicated poor performance during the analysis; or (3) the compound consistently has poor recoveries and concentrations are always reported as estimated (S. Smith, U.S. Geological Survey, oral commun., 2010).

Quality Assurance

Water samples collected for analysis of pharmaceuticals and other organic wastewater compounds require special considerations related to quality-assurance (QA). In this study, QA samples were collected to assess laboratory performance and to help identify potential contamination problems associated with field and/or laboratory methods. These QA samples consisted of equipment blank, field blank, replicate, and matrix spike samples that were collected, processed, and analyzed using methods similar to those used for the environmental samples. Additionally, surrogate compounds were added to all water samples sent to the NWQL to evaluate method performance for individual samples. All quality-assurance results are included in appendixes 1 and 2.

Two field blanks and one equipment blank were collected to ensure that the equipment was adequately cleaned prior to sample collection and that collection and processing did not result in contamination. The blanks were collected by passing laboratory-grade organic-free water through the equipment used for the collection and processing of environmental samples. The two field blanks (one for the churn and one for the groundwater pump) were collected in the field immediately prior to collecting the environmental sample. The equipment blank was collected in the laboratory prior to using any of the sampling and processing equipment (for example, the ISCO pump and churn). Each blank was subject to the same sample processing, handling, and equipment as the environmental samples. The majority of the compounds analyzed were not detected in the blanks collected in this study; only nine compounds (4-nonylphenol monoethoxylate (NP$_1$EO), benzophenone, galaxolide (HHCB), DEET, phenol, TCEP, triclosan, triethyl citrate and triphenyl phosphate) were detected in at least one blank sample (app. 1). Furthermore, concentrations in blanks were typically an order of magnitude lower than concentrations detected in associated environmental samples verifying the general effectiveness of the sampling protocols used in this study. Phenol was the only compound detected in a blank sample above its method reporting level (MRL) (0.51 µg/L, churn blank). Therefore, environmental concentrations of phenol within 10 times the concentration in the churn blank were not reported.

Replicate samples (one canal, one influent, and one effluent) were collected and analyzed to determine the variability in sample collection and processing procedures and to examine the effect these variations can have on evaluating the precision of ambient environmental concentrations. Replicate samples were collected immediately after environmental samples were collected, using the same equipment. The variability of chemical analysis and consistency of sample collection and processing were determined by calculating the relative percent difference (RPD) between the two samples. The RPD is defined as:

$$RPD = (d/\bar{x}) \times 100 \qquad (1)$$

where

d is the difference between the replicate sample and environmental sample, and

$\bar{x}$ is the average concentration of the two samples.

Generally, an RPD of 20 percent or less indicates an acceptable level of precision. However, for very small concentrations near the MRL, the RPD can be higher than 20 percent and still be considered reasonable due to the limits of resolution (Taylor, 1987). Median RPDs for canal, influent, and effluent water were 4.9 percent (ranging from 0.0 to 22.2 percent), 31.5 percent (ranging from 0.8 to 140 percent), and 20.3 percent (ranging from 0.0 to 167 percent), respectively. There were 14 instances (4 canal, 5 influent, and 5 effluent) where a specific compound was detected in only one of the paired replicate samples. Replicate data indicate that there is a higher degree of uncertainty in influent and effluent concentrations compared to canal concentrations.

In an effort to evaluate the impact of matrix effects on overall analytical performance, two environmental samples (one influent and one effluent) were spiked with target analytes. Median matrix spike recoveries for individual compounds in influent and effluent spike samples were 79.7 (ranging from -508 to 757 percent) and 88.3 percent (ranging from 0.0 to 213 percent), respectively (app. 1). Spike recoveries for 76 compounds in the influent spiked sample and 35 compounds in the effluent spiked sample were outside the expected spike recovery range for that particular compound (values shown in red type in app. 1). However, the expected ranges were established using spike results from organic-free water samples and there are no established matrix-specific control limits. Additionally, if sample matrices have environmental concentrations that are higher than the spiked concentrations added to them, which was the case with numerous compounds in both the influent and effluent spiked sample, achieving acceptable spike recoveries might not be possible. Generally, the matrix spike results indicated there were larger matrix effects in influent samples than in effluent samples. Three compounds (chloramphenicol, ibuprofen, and virginiamycin) were not detected in either spiked sample, which suggests the analytical and(or) sampling method did not perform well for the low-level detection of these three compounds.

Surrogate compounds were added to samples at the NWQL to monitor the performance of the analytical processes and extraction steps. The acceptable range for spike recovery established by the NWQL (app. 2) is based on spike recoveries of past sample analyses and is recalculated periodically as more batches of samples are analyzed (S. Abney, U.S. Geological Survey, National Water Quality Laboratory, Denver, Colo., written commun., 2011). Anomalous recovery values that exceed this range can indicate possible analytical problems for individual samples. Alpha-HCH-d6 and diazinon-d10, two surrogate compounds, were added to samples analyzed for pesticides and pesticide degradates. The median apparent recoveries of alpha-HCH-d6 and diazinon-d10 were 91.3 percent (ranging from 74.2 to 175 percent) and 137 percent (ranging from 91.3 to 184 percent), respectively (app. 2). Carbamazepine-d10 and ethylnicotinate-d4, two surrogate compounds, were added to samples analyzed for pharmaceuticals (app. 2). The median apparent recoveries of carbamazepine-d10 and ethylnicotinate-d4 were 20.1 percent (ranging from 7.2 to 119 percent) and 64.5 percent (ranging from 24.3 to 97.2 percent), respectively (app. 2). Bisphenol A-d3, caffeine-C13, decafluorobiphenyl and fluoranthene-d10, four surrogate compounds, were added to samples analyzed for wastewater-indicator compounds (app. 2). The median apparent recoveries of bisphenol A-d3, caffeine-C13,

decafluorobiphenyl and fluoranthene-d10 were 57.3 percent (ranging from 0.0 to 134 percent), 63.2 percent (ranging from 0.0 to 97.4), 42.6 percent (ranging from 14.6 to 77.7 percent), and 58.1 percent (ranging from 10.9 to 85.5 percent), respectively (app. 2). 2,4,6-tribromophenol, 2-fluorobiphenyl, 2-fluorophenol, nitrobenzene-d5, phenol-d5, and terphenyl-d14, six surrogate compounds, were added to samples analyzed for SVOCs (app. 2). The median apparent recoveries of 2,4,6-tribromophenol, 2-fluorobiphenyl, 2-fluorophenol, nitrobenzene-d5, phenol-d5, and terphenyl-d14 were 90.4 percent (ranging from 21.1 to 136), 73.6 percent (ranging from 22.0 to 93.5), 59.1 percent (ranging from 14.8 to 101 percent), 90.8 percent (ranging from 28.6 to 124 percent), 53.7 percent (ranging from 15.1 to 110 percent), and 45.4 percent (ranging from 16.8 to 92.1 percent), respectively (app. 2). Meclocycline, nalidixic acid, and sulfamethazine-$^{13}C_6$ were added to each water sample analyzed for antibiotics. The median apparent recoveries of meclcycline, nalidixic acid, and sulfamethazine-$^{13}C_6$ were 74.4 (ranging from 9.6 to 230 percent), 162 (ranging from 58 to 1,492 percent), and 98.2 (ranging from 11.6 to 240 percent), respectively. The high range in the apparent recoveries for the naxidilic acid indicates that its response was enhanced in the influent and effluent matrices relative to its internal standard.

Estimation of Loads

Daily loads (total mass of compounds discharged in the effluent in one day) were calculated for influent and effluent samples to compare among seasons and to provide information on the estimated quantity of pharmaceuticals and other organic wastewater compounds discharged to the environment from WWTPs in Miami-Dade County. Loads were calculated by multiplying the average flow during each 24-hr sampling period by total concentrations of constituents in the effluent and by a conversion factor (8.34×10^{-9}) to convert million gallons per day and micrograms per liter to pounds per day. Concentrations reported as less than the MRL were censored to 0.0 and not used in the load calculations.

Erroneous load estimates can be caused by two factors. Firstly, many of the concentrations of pharmaceuticals and other organic wastewater compounds were reported as estimated values (E) by the laboratory. Multiplying the concentrations by flow, which in some cases was very large, might result in substantially increasing the effect of analytical error in the reported load estimates. Secondly, several of the estimated loads were calculated using results from filtered water methods (that is, pesticides, pharmaceuticals, antibiotics, and E2), which does not take into account the fraction of compounds in the non-aqueous phase, thus underestimating the total load. Although there might be error in the absolute value of the load estimates, they can provide relative estimates to compare among sites and sampling periods (Sando and others, 2005).

Pharmaceuticals and Other Organic Wastewater Compounds Detected in Wastewater Samples

Influent and effluent samples were collected from the six plants in Miami-Dade County to determine total loads in both the influent and effluent and percent reduction of each target compound through the treatment process. Samples were collected and analyzed from each WWTP on two occasions, once in the dry season and once in the wet season, to determine seasonal differences in concentrations. The influent and effluent samples collected from CD1 in the dry season for the analysis of wastewater-indicator compounds were lost during shipment; therefore, results for those compounds are not available. Sampling dates and average flow during sampling at each plant are listed in table 1. Summarized results (including minimum, median, maximum concentrations detected, and number of detections) for wet-season and dry-season sampling as well as combined sampling are listed in tables 4 and 5.

Pharmaceuticals and Other Organic Wastewater Compounds Detected in 24-Hour Flow-Weighted Influent Composite Samples

Influent samples at each plant contained a complex mixture of organic compounds reflecting the diversity of incoming domestic, municipal, and industrial waste, as well as stormwater runoff. Compounds detected in influent samples included: 20 SVOCs, 12 pesticides, 52 wastewater-indicator compounds, 5 pharmaceuticals, 14 antibiotic, and E2 (table 4; app. 3). Wastewater-indicator compounds accounted for the majority of the total concentration (equal to the sum of all detected concentrations of all analytes) in each effluent sample followed by pharmaceuticals and SVOCs (fig. 3A; app. 4). Pesticides and pesticide degradates, antibiotics, and E2 accounted for less than 2 percent of the total concentration in each influent sample. The total concentrations in influent samples collected in the dry season were generally 10 to 30 percent higher than samples collected in the wet season (fig. 3A; app. 4). The lower total concentrations in influent samples collected in the wet season most likely reflect the higher fraction of stormwater runoff in wastewater during the wet season than during the dry season. Generally, stormwater runoff would be expected to contain smaller concentrations of many of the target compounds (Wilkison and others, 2006), which typically occur more frequently in household and industrial wastewaters that are discharged directly to the sewer system. Estimated loads to each plant ranged from 14.6 to 403.9 pounds per day (lb/d) and were higher in the wet season at five of the six plants (fig. 3B; app. 4). For graphical purposes, E2 was not included in figure 3.

Among the 20 SVOCs detected, diethylhexyl phthalate (DEHP), a widely used plasticizer, showed the largest

concentration range (5.3 to 270 µg/L) and highest median concentration (12 µg/L) (table 4). The three SVOCs 1,4-dichlorobenzene (ranging in concentration from 0.84 to 15 µg/L), 2,4-dichlorophenol (ranging in concentration from 0.10 to 1.6 µg/L), and diethyl phthalate (ranging in concentration from 2.2 to 10 µg/L), were detected in every influent sample. The four compounds 1,2-dichlorobenzene, 2-chlorophenol, benzo[k]fluoranthene, and N-nitrosodiphenylamine, were only detected in dry-season influent samples.

Scattered occurrences of 12 pesticides and pesticides degradates were detected near reporting levels in influent samples; carbaryl was the only pesticide detected at concentrations greater than 0.2 µg/L. Fipronil, a broad spectrum insecticide, and two fipronil degradates (desulfinylfipronil and fipronil sulfide) were the only pesticides detected in more than 50 percent of influent samples collected, with concentrations of the three constituents ranging from 0.005 to 0.15 µg/L (table 4). Generally, pesticides were detected more frequently in influent samples collected in the wet season (summer months) with four pesticides (chlorpyrifos, fipronil sulfone, methidathion, and terbuthylazine) detected in wet-season influent samples only. The higher fraction of stormwater runoff in wastewater during the wet season than during the dry season might contribute to the higher frequency of detection of pesticides during the wet season. The higher detection frequency of pesticides and pesticides degradates in the wet season could reflect higher pesticide usage during warmer months of the wet season.

Among the 52 wastewater-indicator compounds detected, the two biogenic sterols, cholesterol and 3-*beta*-coprostanol, had the highest maximum concentrations (130 and 120 µg/L, respectively) and the highest median concentrations (91 and 79 µg/L, respectively), while the detergent metabolite 4-nonylphenol (NP, total) showed the largest concentration range (3.8 to 120 µg/L). The median concentrations for individual compounds were slightly higher (but not statistically different) in the dry season compared to the wet season (table 4). The six wastewater-indicator compounds (2,2',4,4'-tetrabromodiphenylether, 3,4-dichlorophenyl isocyanate, 4-n-octylphenol, anthraquinone, bromacil, and isoquinoline) were not detected in influent samples (app. 3), although the poor recovery of 3,4-dichlorophenyl isocyanate (0.0 percent) in the influent spike sample (app. 1) affected the probability for its detection.

Among the five pharmaceuticals detected, only one (acetaminophen) was detected in every influent sample (ranging from 18 to 120 µg/L; table 4). Carbamazepine, an anticonvulsant, was detected in all but one influent sample at low concentrations (ranging from 0.029 to 0.23 µg/L). Albuterol, dehydronifedipine, diltiazem, diphenhydramine, and warfarin were not detected in influent samples (app. 3), although the poor recovery of diltiazem (0 percent) and diphenhydramine (0 percent) in the influent spike sample (app. 1) affected the probability for their detection.

Among the 14 antibiotics detected, 6 (ciprofloxacin, erythromycin-H₂O, ofloxacin, sulfamethoxazole, tetracycline and trimethoprim) were detected in every influent sample (ranging from 0.009 to 1.7 µg/L). Sulfamethoxazole, a

Table 4. Summary of concentrations of pharmaceuticals and other organic wastewater compounds detected in 24-hour flow-weighted influent composite samples.

[Concentrations in micrograms per liter. MRL, method reporting level; µg/L, micrograms per liter; NWQL, U.S. Geological Survey National Water Quality Laboratory; CLLE, Continuous liquid-liquid extraction; GC/MS, gas chromatogrpahy/mass spectrometry; OGRL, U.S. Geological Survey Organic Geochemistry Research Laboratory; SPE, solid-phase extraction; HPLC/MS high-performance liquid chromatogrpahy/mass spectrometry; ELISA, enzyme-linked immunosorbent assay; —, not applicable]

Compound	MRL (µg/L)	Wet-season effluent				Dry-season effluent				Combined effluent			
		Minimum concentration detected	Median concentration detected	Maximum concentration detected	Number of detections	Minimum concentration detected	Median concentration detected	Maximum concentration detected	Number of detections	Minimum concentration detected	Median concentration detected	Maximum concentration detected	Number of detections
Semivolatile organic compounds (NWQL, Unfiltered, CLLE, GC/MS)													
1,2-Dichlorobenzene	0.2	—	—	—	—	0.02	0.03	0.05	2	0.02	0.03	0.05	2
1,4-Dichlorobenzene	0.2	0.84	1.6	3.8	6	1.7	2.6	15	6	0.84	2.1	15	12
2,4,6-Trichlorophenol	0.6	0.06	0.09	0.11	4	0.14	0.15	0.36	3	0.06	0.11	0.36	7
2,4-Dichlorophenol	0.39	0.10	0.72	1.0	6	0.74	0.99	1.6	6	0.10	0.87	1.6	12
2,4-Dimethylphenol	0.8	0.31	0.38	0.46	2	0.50	0.50	0.50	1	0.31	0.46	0.50	3
2-Chlorophenol	0.42	—	—	—	—	0.09	0.09	0.09	1	0.09	0.09	0.09	1
4-Chloro-3-methylphenol	0.55	1.1	1.2	1.3	2	0.96	0.96	0.96	1	0.96	1.1	1.3	3
Acenaphthene	0.28	0.1	0.15	0.3	6	0.11	0.16	0.47	5	0.11	0.16	0.47	11
Benz[a]anthracene	0.26	0.0	0.1	0.1	3	0.04	0.04	0.04	1	0.04	0.055	0.079	4
Benzo[b]fluoranthene	0.4	0.1	0.1	0.1	1	0.08	0.08	0.08	1	0.08	0.096	0.12	2
Benzo[k]fluoranthene	0.4	—	—	—	—	0.03	0.03	0.032	1	0.03	0.032	0.032	1
Butylbenzyl phthalate	1.8	1.1	1.4	3	6	1.9	4.4	6.8	2	1.1	1.6	6.8	8
Chrysene	0.33	0.06	0.09	0.10	3	0.043	0.063	1.2	4	0.043	0.064	1.2	7
Diethyl phthalate	0.62	2.2	3.8	7.7	6	4.6	8.7	10	6	2.2	7.0	10	12
Diethylhexyl phthalate (DEHP)	1.3	5.3	10	32	6	11	12	270	5	5.3	12	270	11
Di-n-butyl phthalate	1	0.70	0.7	1	5	2.0	2.0	2.0	1	0.70	0.83	2.0	6
Di-n-octyl phthalate	0.6	0.16	0.27	0.51	5	0.59	0.74	0.89	2	0.16	0.47	0.89	7
Fluorene	0.33	0.095	0	0	5	0.15	0.17	0.34	3	0.095	0.15	0.34	8
N-Nitrosodiphenylamine	0.4	—	—	—	—	0.25	0.25	0.25	1	0.25	0.25	0.25	1
Pentachlorophenol	0.6	0.18	0.18	0.18	1	0.4	0.4	0.4	1	0.18	0.29	0.4	2
Total					67				53				120

Table 4. Summary of concentrations of pharmaceuticals and other organic wastewater compounds detected in 24-hour flow-weighted influent composite samples.—Continued

[Concentrations in micrograms per liter. MRL, method reporting level; µg/L, micrograms per liter; NWQL, U.S. Geological Survey National Water Quality Laboratory; CLLE, Continuous liquid-liquid extraction; GC/MS, gas chromatograpahy/mass spectrometry; OGRL, U.S. Geological Survey Organic Geochemistry Research Laboratory; SPE, solid-phase extraction; HPLC/MS high-performance liquid chromatogrpahy/mass spectrometry; ELISA, enzyme-linked immunosorbent assay; —, not applicable]

Compound	MRL (µg/L)	Wet-season effluent				Dry-season effluent				Combined effluent			
		Minimum concentration detected	Median concentration detected	Maximum concentration detected	Number of detections	Minimum concentration detected	Median concentration detected	Maximum concentration detected	Number of detections	Minimum concentration detected	Median concentration detected	Maximum concentration detected	Number of detections
Pesticides and pesticide degradates (NWQL, Filtered, SPE, GC/MS)													
1-Napthol	0.04	0.14	0.14	0.14	1	0.05	0.12	0.18	2	0.05	0.14	0.18	3
3,4-Dichloroaniline	0.004	0.05	0.11	0.13	4	0.18	0.18	0.18	1	0.05	0.12	0.18	5
Atrazine	0.007	0.02	0.04	0.04	3	0.04	0.04	0.04	1	0.02	0.04	0.04	4
Carbaryl	0.2	0.04	0.11	0.56	4	0.25	0.25	0.25	1	0.04	0.18	0.56	5
Chlorpyrifos	0.01	0.01	0.01	0.02	3	—	—	—	—	0.01	0.01	0.02	3
Desulfinylfipronil	0.012	0.005	0.01	0.007	4	0.006	0.006	0.008	4	0.005	0.006	0.008	8
Fipronil	0.04	0.036	0.05	0.08	6	0.024	0.053	0.15	6	0.024	0.048	0.15	12
Fipronil sulfide	0.013	0.008	0.008	0.01	6	0.006	0.009	0.014	6	0.006	0.008	0.014	11
Fipronil sulfone	0.024	0.01	0.01	0.01	1	—	—	—	—	0.01	0.01	0.01	1
Hexazinone	0.008	0.023	0.03	0.034	4	0.028	0.031	0.033	2	0.023	0.029	0.034	6
Methidathion	0.006	0.11	0.11	0.11	1	—	—	—	—	0.11	0.11	0.11	1
Terbuthylazine	0.006	0.12	0.12	0.12	1	—	—	—	—	0.12	0.12	0.12	1
Total					38				22				60
Wastewater-indicator compounds (NWQL, Unfiltered, CLLE, GC/MS)													
1,7-Dimethylxanthine	0.1	1.2	1.3	7.7	6	1.2	9.1	11	6	1.2	6.4	11	12
1-Methylnaphthalene	0.2	0.064	0.14	0.44	6	0.050	0.12	0.42	5	0.050	0.12	0.44	11
2,6-Dimethylnaphthalene	0.2	0.049	0.089	0.48	6	0.019	0.11	0.56	5	0.019	0.11	0.56	11
2-Methylnaphthalene	0.2	0.16	0.30	0.79	6	0.11	0.27	0.76	5	0.11	0.27	0.79	11
3-*beta*-Coprostanol	1.6	50	75	100	6	63.3	87	120	6	50	79	120	11
3-Methyl-1H-indole (skatol)	0.2	0.46	1.1	1.5	6	0.34	1.1	1.8	5	0.34	1.1	1.8	11
3-*tert*-Butyl-4-hydroxyanisole (BHA)	0.2	0.11	0.12	0.18	5	0.059	0.27	0.32	5	0.059	0.14	0.32	10
4-Cumylphenol	0.2	0.085	0.089	0.092	2	0.19	0.19	0.19	1	0.085	0.092	0.19	3
4-Nonylphenol (total, NP)	1.6	5.2	24	46	6	3.8	36	120	5	3.8	27	120	11
4-Nonylphenol diethoxylate (NP$_2$EO)	3.2	1.6	5.5	8.2	3	4.3	21	37	2	1.6	5.5	37	5

Table 4. Summary of concentrations of pharmaceuticals and other organic wastewater compounds detected in 24-hour flow-weighted influent composite samples.—Continued

[Concentrations in micrograms per liter. MRL, method reporting level; μg/L, micrograms per liter; NWQL, U.S. Geological Survey National Water Quality Laboratory; CLLE, Continuous liquid-liquid extraction, GC/MS, gas chromatogrpahy/mass spectrometry; OGRL, U.S. Geological Survey Organic Geochemistry Research Laboratory; SPE, solid-phase extraction; HPLC/MS high-performance liquid chromatogrpahy/mass spectrometry; ELISA, enzyme-linked immunosorbent assay; —, not applicable]

Compound	MRL (μg/L)	Wet-season effluent				Dry-season effluent				Combined effluent			
		Minimum concentration detected	Median concentration detected	Maximum concentration detected	Number of detections	Minimum concentration detected	Median concentration detected	Maximum concentration detected	Number of detections	Minimum concentration detected	Median concentration detected	Maximum concentration detected	Number of detections
4-Nonylphenol monoethoxylate (NP$_1$EO)	1.6	4.0	15	16	3	3.5	6.7	26	3	3.5	11	26	6
4-Octylphenol diethoxylate (OP$_2$EO)	0.5	3.3	4.5	6.7	6	12	33	48	5	1.2	5.7	11	11
4-Octylphenol monoethoxylate (OP$_1$EO)	1	1.3	3.2	5.5	4	2.6	2.9	4.4	4	3.3	6.7	48	8
4-*tert*-Octylphenol (OP)	0.4	1.2	5.2	9.1	6	3.7	6.8	11	4	1.3	3.0	5.5	10
5-Methyl-1H-benzotriazole	1.6	0.51	0.51	0.51	1	0.71	0.71	0.71	1	0.51	0.61	0.71	2
Acetophenone	0.4	0.48	0.58	0.89	6	0.69	1.2	1.7	4	0.48	0.75	1.7	10
Anthracene	0.2	0.028	0.038	0.079	3	0.034	0.039	0.073	4	0.028	0.039	0.079	7
Benzo[a]pyrene	0.2	0.01	0.02	0.10	5	0.026	0.026	0.026	1	0.01	0.025	0.10	6
Benzophenone	0.2	0.49	0.55	1.4	6	0.24	0.89	1.4	5	0.24	0.57	1.4	11
beta-Sitosterol	1.6	7.6	8.7	29	6	7.8	9.0	14	5	7.6	9.0	29	11
beta-Stigmastanol	1.7	1.4	2.4	7.7	6	2.1	3.0	4.6	5	1.4	2.8	7.7	11
Bisphenol A	0.4	1.1	4.3	6.7	6	0.54	11	21.8	2	0.54	4.2	22	8
Bromoform	0.2	0.059	0.059	0.059	1	—	—	—	—	0.059	0.059	0.059	1
Caffeine	0.06	22	27	38	6	32	38	50	6	22	34	50	12
Camphor	0.2	1.2	2.3	3.4	6	0.61	2.8	3.4	5	0.61	2.7	3.4	11
Carbazole	0.2	0.14	0.14	0.14	1	0.053	0.053	0.053	1	0.053	0.098	0.14	2
Cholesterol	1.6	74	90	130	6	56	93	130	5	56	91	130	11
Cotinine	0.8	0.47	0.55	0.65	3	0.35	0.35	0.35	1	0.35	0.51	0.65	4
Fluoranthene	0.2	0.059	0.078	0.19	6	0.051	0.061	0.070	2	0.051	0.066	0.19	8
Galaxolide (HHCB)	0.2	2.9	4.2	7.4	6	1.7	6.1	7.1	5	1.7	4.3	7.4	11
Indole	0.2	0.066	2.1	4.1	6	0.16	1.9	2.4	3	0.066	2.0	4.1	9
Isoborneol	0.2	0.46	0.98	1.2	6	0.30	1.2	1.4	5	0.30	1.1	1.4	11
Isophorone	0.2	0.52	0.52	0.52	1	—	—	—	—	0.52	0.52	0.52	1
Isopropylbenzene (cumene)	0.2	0.006	0.065	0.11	4	0.057	0.07	0.083	2	0.006	0.07	0.11	6
Limonene	0.2	1.0	3.4	7.5	6	1.1	2.3	9.8	5	1.0	2.3	9.8	11

Table 4. Summary of concentrations of pharmaceuticals and other organic wastewater compounds detected in 24-hour flow-weighted influent composite samples.—Continued

[Concentrations in micrograms per liter. MRL, method reporting level; µg/L, micrograms per liter; NWQL, U.S. Geological Survey National Water Quality Laboratory; CLLE, Continuous liquid-liquid extraction; GC/MS, gas chromatograpahy/mass spectrometry; OGRL, U.S. Geological Survey Organic Geochemistry Research Laboratory; SPE, solid-phase extraction; HPLC/MS high-performance liquid chromatorgrahy/mass spectrometry; ELISA, enzyme-linked immunosorbent assay; —, not applicable]

Compound	MRL (µg/L)	Wet-season effluent				Dry-season effluent				Combined effluent			
		Minimum concentration detected	Median concentration detected	Maximum concentration detected	Number of detections	Minimum concentration detected	Median concentration detected	Maximum concentration detected	Number of detections	Minimum concentration detected	Median concentration detected	Maximum concentration detected	Number of detections
Menthol	0.2	7.4	11	16	6	4.1	14	19	5	4.1	13	19	11
Methyl salicylate	0.2	0.20	0.46	0.74	6	0.25	0.69	0.70	3	0.20	0.47	0.74	9
N,N-Diethyl-meta-toluamide (DEET)	0.2	0.79	1.1	1.6	5	0.19	0.50	0.64	5	0.19	0.72	1.6	10
Naphthalene	0.2	0.23	0.32	0.60	6	0.11	0.31	0.60	5	0.11	0.31	0.60	11
para-Cresol	0.2	0.06	14	46	6	5.1	29	65	3	0.06	21	65	9
Phenanthrene	0.2	0.14	0.18	0.29	6	0.13	0.18	0.30	5	0.13	0.18	0.30	11
Phenol	0.2	6.8	12	15	3	16	19	21	2	6.8	15	21	5
Pyrene	0.2	0.017	0.022	0.026	2	—	—	—	—	0.017	0.022	0.026	2
Tetrachloroethylene	0.4	0.008	0.21	0.32	6	0.048	0.13	0.53	5	0.008	0.21	0.53	11
Tonalide (AHTN)	0.2	0.32	0.60	0.78	4	0.17	0.70	0.78	3	0.17	0.66	0.78	7
Tri(2-butoxyethyl) phosphate (TBEP)	0.2	4.2	6.0	15	6	3.3	8.4	11	5	3.3	6.4	15	11
Tri(2-chloroethyl) phosphate (TLEP)	0.2	0.16	0.17	0.20	4	0.12	0.12	0.12	1	0.12	0.16	0.20	5
Tri(dichloroisopropyl) phosphate (TDIP)	0.2	0.17	0.28	0.35	6	0.16	0.36	0.45	4	0.16	0.31	0.45	10
Tributyl phosphate (TBP)	0.2	0.11	0.17	0.35	4	0.29	0.29	0.29	1	0.11	0.19	0.35	5
Triclosan	0.2	2.4	3.5	5.6	6	1.5	3.8	7.8	5	1.5	3.6	7.8	11
Triethyl citrate (ethyl citrate)	0.2	0.39	0.48	1.1	6	0.26	0.81	0.98	5	0.26	0.56	1.1	11
Triphenyl phosphate	0.2	0.07	0.09	0.13	6	0.077	0.14	0.15	4	0.07	0.11	0.15	10
Total					244				177				421
Pharmaceutical compounds (NWQL, Filtered, SPE, HPLC/MS)													
Acetaminophen	0.08	18	26	45	6	29	40	122	6	18	35	120	12
Carbamazepine	0.005	0.029	0.15	0.23	6	0.084	0.14	0.19	5	0.029	0.15	0.23	11
Codeine	0.04	0.004	0.007	0.011	2	0.009	0.027	0.028	3	0.004	0.011	0.028	5
Ibuprofen	0.05	1.6	6.6	9.5	3	0.16	5.2	13	6	0.16	5.2	13	9
Thiabendazole	0.06	0.005	0.005	0.005	1	—	—	—	—	0.005	0.005	0.005	1
Total					18				20				38

Table 4. Summary of concentrations of pharmaceuticals and other organic wastewater compounds detected in 24-hour flow-weighted influent composite samples.—Continued

[Concentrations in micrograms per liter. MRL, method reporting level; μg/L, micrograms per liter; NWQL, U.S. Geological Survey National Water Quality Laboratory; CLLE, Continuous liquid-liquid extraction; GC/MS, gas chromatograpahy/mass spectrometry; OGRL, U.S. Geological Survey Organic Geochemistry Research Laboratory; SPE, solid-phase extraction; HPLC/MS high-performance liquid chromatograpahy/mass spectrometry; ELISA, enzyme-linked immunosorbent assay; —, not applicable]

Compound	MRL (μg/L)	Wet-season effluent				Dry-season effluent				Combined effluent			
		Minimum concentration detected	Median concentration detected	Maximum concentration detected	Number of detections	Minimum concentration detected	Median concentration detected	Maximum concentration detected	Number of detections	Minimum concentration detected	Median concentration detected	Maximum concentration detected	Number of detections
Antibiotics (OGRL, Filtered, SPE, LC/MS)													
Azithromycin	0.005	0.026	0.74	1.6	5	0.039	0.13	0.38	4	0.026	0.18	1.6	9
Ciprofloxacin	0.005	0.54	1.0	1.4	6	0.25	0.62	0.95	6	0.25	0.83	1.4	12
Doxycycline	0.01	0.03	0.041	0.052	2	0.051	0.087	0.12	2	0.03	0.052	0.12	4
epi-Tetracycline	0.01	0.012	0.012	0.012	1	0.039	0.039	0.039	1	0.012	0.026	0.039	2
Erythromycin	0.008	0.008	0.009	0.01	3	—	—	—	—	0.008	0.009	0.01	3
Erythromycin-H₂O	0.008	0.009	0.02	0.19	6	0.074	0.18	0.22	6	0.009	0.15	0.22	12
Lincomycin	0.005	0.007	0.007	0.007	1	0.011	0.011	0.011	1	0.007	0.009	0.011	2
Norfloxacin	0.005	0.035	0.035	0.035	1	0.011	0.011	0.011	1	0.007	0.009	0.011	2
Ofloxacin	0.005	0.24	0.46	0.60	6	0.16	0.97	1.7	6	0.16	0.57	1.7	12
Sulfadiazine	0.1	0.063	0.084	0.11	2	1.0	1.0	1.0	1	0.063	0.11	1.0	3
Sulfamethoxazole	0.005	0.3	1.3	1.7	6	0.029	0.27	1.6	6	0.029	1.1	1.7	12
Sulfathiazole	0.05	—	—	—	—	0.073	0.073	0.073	1	0.073	0.073	0.073	1
Tetracycline	0.01	0.05	0.10	0.17	6	0.06	0.12	0.38	6	0.05	0.11	0.38	12
Trimethoprim	0.005	0.091	0.33	0.69	6	0.20	0.46	0.66	6	0.091	0.38	0.69	12
Total					51				47				98
Hormone (OGRL, Filtered, ELISA)													
17-*beta*-Estradiol (E2)	0.0015	0.016	0.040	0.076	6	0.011	0.052	0.22	5	0.011	0.042	0.22	11

Table 5. Summary of concentrations of pharmaceuticals and other organic wastewater compounds detected in 24-hour flow-weighted effluent composite samples.

[Concentrations in micrograms per liter. MRL, method reporting level; µg/L, micrograms per liter; NWQL, U.S. Geological Survey National Water Quality Laboratory; CLLE, Continuous liquid–liquid extraction; GC/MS, gas chromatograpahy/mass spectrometry; OGRL, U.S. Geological Survey Organic Geochemistry Research Laboratory; SPE, solid-phase extraction; HPLC/MS high-performance liquid chromatograpahy/mass spectrometry; ELISA, enzyme-linked immunosorbent assay; —, not applicable]

Compound	MRL (µg/L)	Wet-season effluent				Dry-season effluent				Combined effluent			
		Minimum concentration detected	Median concentration detected	Maximum concentration detected	Number of detections	Minimum concentration detected	Median concentration detected	Maximum concentration detected	Number of detections	Minimum concentration detected	Median concentration detected	Maximum concentration detected	Number of detections
Semivolatile organic compounds (NWQL, unfiltered, CLLE, GC/MS)													
1,2,4-Trichlorobenzene	0.2	0.01	0.024	0.038	2	—	—	—	—	0.01	0.02	0.04	2
1,2-Dichlorobenzene	0.2	0.026	0.032	0.067	4	0.015	0.035	0.041	3	0.015	0.03	0.07	7
1,3-Dichlorobenzene	0.2	0.012	0.025	0.026	3	0.006	0.006	0.006	1	0.006	0.018	0.03	4
1,4-Dichlorobenzene	0.2	0.12	0.92	2.3	6	0.25	0.78	1.2	6	0.12	0.82	2.3	12
2,4,6-Trichlorophenol	0.6	0.021	0.074	0.13	6	0.044	0.12	0.14	6	0.021	0.09	0.14	12
2,4-Dichlorophenol	0.39	0.05	0.17	0.45	6	0.091	0.16	0.19	5	0.05	0.16	0.45	11
2,4-Dimethylphenol	0.8	0.02	0.039	0.18	4	0.085	0.14	0.26	4	0.02	0.11	0.26	8
2-Chloronaphthalene	0.2	0.033	0.033	0.033	1	—	—	—	—	0.033	0.033	0.033	1
2-Nitrophenol	0.4	0.16	0.17	0.17	2	—	—	—	—	0.16	0.17	0.17	2
Acenaphthene	0.28	0.035	0.035	0.035	1	0.02	0.02	0.02	1	0.02	0.027	0.035	2
Butylbenzyl phthalate	1.8	0.65	0.65	0.65	1	—	—	—	—	0.65	0.65	0.65	1
Diethylhexyl phthalate (DEHP)	1.3	0.5	0.6	0.7	2	1	1	1	1	0.5	0.7	1	3
Dimethyl phthalate	0.4	0.053	0.17	0.3	2	0.073	0.073	0.073	1	0.053	0.073	0.3	3
Di-n-butyl phthalate	1	0.2	0.2	0.2	1	—	—	—	—	0.2	0.2	0.2	1
Di-n-octyl phthalate	0.6	0.09	0.09	0.09	1	—	—	—	—	0.09	0.09	0.09	1
Fluorene	0.33	0.01	0.014	0.015	3	0.013	0.014	0.016	3	0.01	0.014	0.016	6
Nitrobenzene	0.2	0.022	0.035	0.047	2	0.045	0.045	0.045	1	0.022	0.045	0.047	3
N-Nitrosodimethyl-amine (NDMA)	0.2	0.035	0.046	0.057	2	0.016	0.016	0.016	1	0.016	0.035	0.057	3
Pentachlorophenol	0.6	—	—	—	—	0.23	0.23	0.23	1	0.23	0.23	0.23	1
Total					49				34				83
Pesticides and pesticide degradates (NWQL, filtered, SPE, GC/MS)													
1-Naphthol	0.04	—	—	—	—	0.012	0.015	0.017	2	0.012	0.015	0.017	2
2-Chloro-4-isopropyl-amino-6-amino-s-triazine (CIAT)	0.014	0.018	0.018	0.018	1	0.021	0.021	0.021	1	0.018	0.02	0.021	2

Table 5. Summary of concentrations of pharmaceuticals and other organic wastewater compounds detected in 24-hour flow-weighted effluent composite samples.—Continued

[Concentrations in micrograms per liter. MRL, method reporting level; µg/L, micrograms per liter; NWQL, U.S. Geological Survey National Water Quality Laboratory; CLLE, Continuous liquid-liquid extraction; GC/MS, gas chromatography/mass spectrometry; OGRL, U.S. Geological Survey Organic Geochemistry Research Laboratory; SPE, solid-phase extraction; HPLC/MS high-performance liquid chromatography/mass spectrometry; ELISA, enzyme-linked immunosorbent assay; —, not applicable]

Compound	MRL (µg/L)	Wet-season effluent				Dry-season effluent				Combined effluent			
		Minimum concentration detected	Median concentration detected	Maximum concentration detected	Number of detections	Minimum concentration detected	Median concentration detected	Maximum concentration detected	Number of detections	Minimum concentration detected	Median concentration detected	Maximum concentration detected	Number of detections
3,4-Dichloroaniline	0.004	0.087	0.095	0.24	6	0.046	0.14	0.21	6	0.046	0.1	0.24	12
Atrazine	0.007	0.018	0.021	0.039	6	0.026	0.032	0.048	5	0.018	0.029	0.048	11
Carbaryl	0.2	0.021	0.041	0.11	6	0.052	0.06	0.29	5	0.021	0.058	0.29	11
Chlorpyrifos	0.01	0.006	0.009	0.013	5	0.007	0.007	0.007	1	0.006	0.008	0.013	6
Desulfinylfipronil	0.012	0.005	0.006	0.012	5	0.005	0.005	0.013	5	0.005	0.005	0.013	10
Fipronil	0.04	0.016	0.043	0.079	6	0.027	0.06	0.11	6	0.016	0.054	0.11	12
Fipronil sulfide	0.013	0.007	0.01	0.011	6	0.006	0.01	0.019	6	0.006	0.01	0.019	12
Fipronil sulfone	0.024	—	—	—	—	0.009	0.01	0.011	3	0.009	0.01	0.011	3
Hexazinone	0.008	0.017	0.027	0.035	4	0.03	0.036	0.042	5	0.017	0.032	0.042	9
Metribuzin	0.016	—	—	—	—	0.022	0.022	0.022	1	0.022	0.022	0.022	1
Terbuthylazine	0.006	0.06	0.06	0.06	1	—	—	—	—	0.06	0.06	0.06	1
Total					46				46				92
Wastewater-indicator compounds (NWQL, unfiltered, CLLE, GC/MS)													
1-Methylnaphthalene	0.2	—	—	—	—	0.008	0.008	0.008	1	0.008	0.008	0.008	1
2,6-Dimethylnaphthalene	0.2	0.006	0.01	0.013	2	—	—	—	—	0.006	0.01	0.013	2
2-Methylnaphthalene	0.2	—	—	—	—	0.011	0.011	0.011	1	0.011	0.011	0.011	1
Anthracene	0.2	0.013	0.013	0.013	1	0.011	0.011	0.011	1	0.011	0.012	0.013	2
Anthraquinone	0.2	0.03	0.034	0.038	2	0.022	0.022	0.022	1	0.022	0.03	0.038	3
Benzo[a]pyrene	0.2	0.013	0.013	0.013	1	—	—	—	—	0.013	0.013	0.013	1
Fluoranthene	0.2	0.012	0.018	0.022	4	0.01	0.02	0.025	3	0.01	0.019	0.025	7
Isophorone	0.2	0.008	0.029	0.039	6	—	—	—	—	0.008	0.029	0.039	6
Methyl salicylate	0.2	0.02	0.02	0.02	1	—	—	—	—	0.02	0.02	0.02	1
Naphthalene	0.2	0.013	0.013	0.013	1	0.018	0.018	0.018	1	0.013	0.015	0.018	2
Phenanthrene	0.2	0.013	0.015	0.016	2	0.013	0.02	0.02	1	0.013	0.016	0.02	3
Pyrene	0.2	0.014	0.015	0.018	5	0.014	0.02	0.025	4	0.014	0.016	0.025	9
3,4-Dichlorophenyl isocyanate	1.6	0.053	0.25	0.35	6	0.046	0.17	0.22	5	0.046	0.21	0.35	11

Table 5. Summary of concentrations of pharmaceuticals and other organic wastewater compounds detected in 24-hour flow-weighted effluent composite samples.—Continued

[Concentrations in micrograms per liter. MRL, method reporting level; µg/L, micrograms per liter; NWQL, U.S. Geological Survey National Water Quality Laboratory; CLLE, Continuous liquid-liquid extraction; GC/MS, gas chromatograpy/mass spectrometry; OGRL, U.S. Geological Survey Organic Geochemistry Research Laboratory; SPE, solid-phase extraction; HPLC/MS high-performance liquid chromatograpy/mass spectrometry; ELISA, enzyme-linked immunosorbent assay; —, not applicable]

Compound	MRL (µg/L)	Wet-season effluent				Dry-season effluent				Combined effluent			
		Minimum concentration detected	Median concentration detected	Maximum concentration detected	Number of detections	Minimum concentration detected	Median concentration detected	Maximum concentration detected	Number of detections	Minimum concentration detected	Median concentration detected	Maximum concentration detected	Number of detections
3-methyl-1H-indole (skatol)	0.2	0.004	0.036	0.062	4	0.004	0.018	0.023	3	0.004	0.018	0.062	7
4-Octylphenol diethoxylate (OP$_2$EO)	0.5	0.056	0.9	1.3	6	0.82	1.2	2	4	0.043	0.24	0.41	10
Acetophenone	0.4	0.16	0.28	0.58	4	—	—	—	—	0.16	0.28	0.58	4
Benzophenone	0.2	0.1	0.25	0.56	6	0.16	0.37	0.51	4	0.1	0.34	0.56	10
beta-Stigmastanol	1.7	0.56	0.56	0.56	1	—	—	—	—	0.56	0.56	0.56	1
Camphor	0.2	0.027	0.041	0.055	2	0.1	0.1	0.1	1	0.027	0.055	0.1	3
Cotinine	0.8	0.049	0.09	0.22	3	0.073	0.14	0.44	3	0.049	0.11	0.44	6
3-tert-Butyl-4-hydroxyanisole (BHA)	0.2	0.29	0.29	0.29	1	0.12	0.15	0.31	4	0.12	0.18	0.31	5
Indole	0.2	0.056	0.056	0.056	2	—	—	—	—	0.056	0.056	0.056	2
Limonene	0.2	—	—	—	—	0.087	0.087	0.087	1	0.087	0.087	0.087	1
Menthol	0.2	0.032	0.12	0.37	4	—	—	—	—	0.032	0.12	0.37	4
N,N-Diethyl-meta-toluamide (DEET)	0.2	0.04	0.22	0.47	6	0.075	0.19	0.29	3	0.04	0.19	0.47	9
para-Cresol	0.2	0.056	0.19	0.44	4	0.25	0.27	0.29	2	0.056	0.27	0.44	6
Tetrachloroethylene	0.4	0.086	0.1	0.21	5	0.12	0.18	0.22	4	0.086	0.14	0.22	9
Tonalide (AHTN)	0.2	0.045	0.13	0.33	6	0.11	0.21	0.32	4	0.045	0.15	0.33	10
Tri(2-chloroethyl) phosphate	0.2	0.096	0.2	0.26	6	0.2	0.27	0.3	4	0.096	0.23	0.3	10
Tri(dichloroisopropyl) phosphate	0.2	0.13	0.32	0.52	6	0.25	0.42	0.51	4	0.13	0.35	0.52	10
Tributyl phosphate	0.2	0.097	0.17	0.24	6	0.16	0.21	0.43	4	0.097	0.18	0.43	10
Triphenyl phosphate	0.2	0.03	0.044	0.09	6	0.058	0.096	0.15	4	0.03	0.067	0.15	10
1,7-Dimethylxanthine	0.1	0.45	0.45	0.45	1	0.14	0.53	1.3	4	0.14	0.45	1.3	5
3-beta-Coprostanol	1.6	2.6	3.1	4.2	3	0.24	2.3	4.4	5	0.24	2.7	4.4	8
4-Nonylphenol (total, NP)	1.6	0.65	1.4	2.4	5	1.8	2.1	4.7	4	0.65	2	4.7	9

Table 5. Summary of concentrations of pharmaceuticals and other organic wastewater compounds detected in 24-hour flow-weighted effluent composite samples.—Continued

[Concentrations in micrograms per liter. MRL, method reporting level; µg/L, micrograms per liter; NWQL, U.S. Geological Survey National Water Quality Laboratory; CLLE, Continuous liquid-liquid extraction; GC/MS, gas chromatograpahy/mass spectrometry; OGRL, U.S. Geological Survey Organic Geochemistry Research Laboratory; SPE, solid-phase extraction; HPLC/MS high-performance liquid chromatograpahy/mass spectrometry; ELISA, enzyme-linked immunosorbent assay; —, not applicable]

Compound	MRL (µg/L)	Wet-season effluent				Dry-season effluent				Combined effluent			
		Minimum concentration detected	Median concentration detected	Maximum concentration detected	Number of detections	Minimum concentration detected	Median concentration detected	Maximum concentration detected	Number of detections	Minimum concentration detected	Median concentration detected	Maximum concentration detected	Number of detections
4-Nonylphenol diethoxylate (NP$_2$EO)	3.2	0.82	1.5	4.4	4	1.4	2.6	5.2	4	0.82	1.9	5.2	8
4-Nonylphenol monoethoxylate (NP$_1$EO)	1.6	0.53	0.59	0.94	3	0.85	1.4	2	4	0.53	0.94	2	7
4-Octylphenol monoethoxylate (OP$_1$EO)	1	0.62	0.67	0.91	5	1.4	1.4	1.4	1	0.056	1	2	6
4-*tert*-Octylphenol (OP)	0.4	0.043	0.24	0.41	5	—	—	—	—	0.62	0.67	1.4	5
5-Methyl-1H-benzotriazole	1.6	0.1	0.37	0.59	5	0.35	0.52	1.2	3	0.1	0.38	1.2	8
beta-Sitosterol	1.6	1.1	1.5	2.3	3	0.44	0.66	1.2	5	0.44	0.95	2.3	8
Bisphenol A	0.4	0.093	0.21	0.35	5	0.25	0.29	0.39	3	0.093	0.29	2.4	8
Bromoform	0.2	0.013	0.19	0.93	4	0.014	0.016	0.018	2	0.013	0.02	0.93	6
Caffeine	0.06	0.053	0.4	7	6	0.1	1.1	25	6	0.053	0.73	25	12
Cholesterol	1.6	2.7	3.5	4.1	4	0.39	2.4	5.4	5	0.39	2.9	5.4	9
Galaxolide (HHCB)	0.2	0.8	2	4.3	6	1	2.6	4	5	0.8	2	4.3	11
Tri(2-butoxyethyl) phosphate	0.2	1.3	3.3	12	6	1.2	3.1	5.5	5	1.2	3.1	12	11
Triclosan	0.2	0.34	0.57	0.8	3	0.37	0.69	1	3	0.34	0.63	1	6
Triethyl citrate (ethyl citrate)	0.2	0.12	0.36	0.67	6	0.4	0.47	0.67	4	0.12	0.42	0.67	10
Total					183				128				311
Pharmaceutical compounds (NWQL, filtered, SPE, HPLC/MS)													
Acetaminophen	0.08	0.021	0.04	0.059	2	—	—	—	—	0.021	0.04	0.059	2
Carbamazepine	0.005	0.12	0.16	0.19	6	0.13	0.18	0.25	6	0.12	0.17	0.25	12
Codeine	0.04	0.003	0.032	0.051	5	0.03	0.045	0.055	4	0.003	0.033	0.055	9
Dehydronifedipine	0.08	0.006	0.006	0.006	1	0.003	0.007	0.012	5	0.003	0.007	0.012	6
Diltiazem	0.08	0.019	0.032	0.044	2	0.013	0.021	0.049	6	0.013	0.021	0.049	8

Table 5. Summary of concentrations of pharmaceuticals and other organic wastewater compounds detected in 24-hour flow-weighted effluent composite samples.—Continued

[Concentrations in micrograms per liter. MRL, method reporting level; µg/L, micrograms per liter; NWQL, U.S. Geological Survey National Water Quality Laboratory; CLLE, Continuous liquid–liquid extraction; GC/MS, gas chromatograpahy/mass spectrometry; OGRL, U.S. Geological Survey Organic Geochemistry Research Laboratory; SPE, solid-phase extraction; HPLC/MS high-performance liquid chromatogrpahy/mass spectrometry; ELISA, enzyme-linked immunosorbent assay; —, not applicable]

Compound	MRL (µg/L)	Wet-season effluent				Dry-season effluent				Combined effluent			
		Minimum concentration detected	Median concentration detected	Maximum concentration detected	Number of detections	Minimum concentration detected	Median concentration detected	Maximum concentration detected	Number of detections	Minimum concentration detected	Median concentration detected	Maximum concentration detected	Number of detections
Diphenhydramine	0.04	0.011	0.057	0.18	5	0.024	0.066	0.12	6	0.011	0.06	0.18	11
Ibuprofen	0.05	—	—	—	—	0.21	0.52	1.3	4	0.21	0.52	1.3	4
Total					21				31				52
Antibiotics (OGRL, filtered, SPE, LC/MS)													
Azithromycin	0.005	0.15	0.26	0.56	6	0.2	0.46	0.59	6	0.15	0.36	0.59	12
Ciprofloxacin	0.005	0.068	0.28	0.58	6	0.097	0.66	1.1	6	0.068	0.47	1.1	12
Doxycycline	0.01	0.019	0.019	0.019	1	—	—	—	—	0.019	0.019	0.019	1
Erythromycin	0.008	0.012	0.018	0.02	6	0.01	0.013	0.016	2	0.01	0.016	0.02	8
Erythromycin-H_2O	0.008	0.019	0.057	0.077	6	0.082	0.1	0.13	5	0.019	0.077	0.13	11
Ofloxacin	0.005	0.3	0.37	0.67	6	0.34	0.69	1.2	6	0.3	0.56	1.2	12
Sulfadiazine	0.1	0.022	0.027	0.032	2	—	—	—	—	0.022	0.027	0.032	2
Sulfamethoxazole	0.005	0.022	0.29	0.62	5	0.005	0.15	1.9	6	0.005	0.22	1.9	11
Tetracycline	0.01	0.01	0.032	0.048	5	0.016	0.031	0.098	5	0.01	0.032	0.098	10
Total					52				42				94
Trimethoprim	0.005	0.03	0.28	0.34	6	0.094	0.4	1.1	6	0.03	0.32	1.1	12
Tylosin	0.008	0.017	0.022	0.023	3	—	—	—	—	0.017	0.022	0.023	3
Hormone (OGRL, filtered, ELISA)													
17-*beta*-Estradiol (E2)	0.0015	0.011	0.021	0.08	6	0.0097	0.015	0.016	4	0.0097	0.018	0.08	10

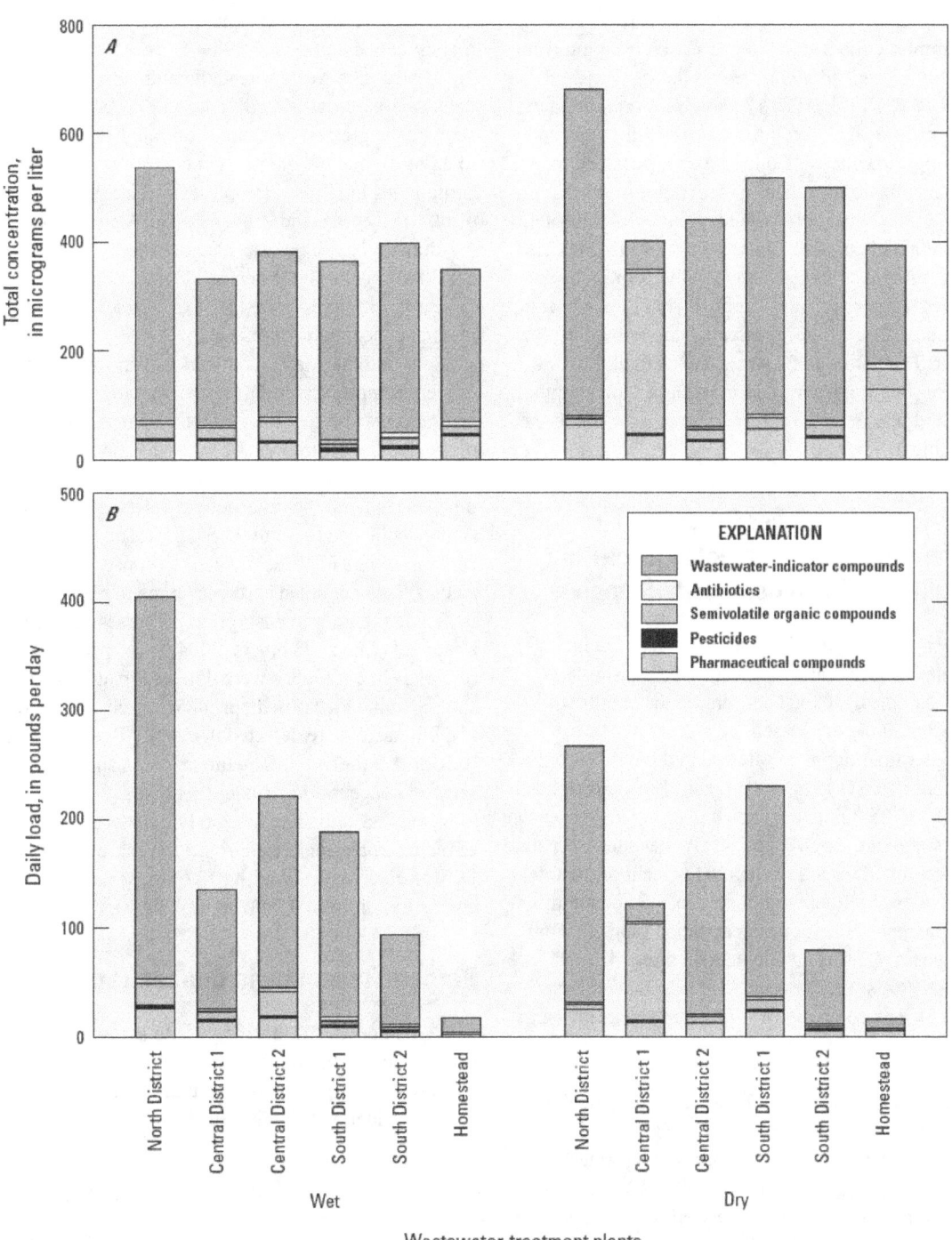

Figure 3. Total concentration *A*, and daily loads *B*, of pharmaceuticals and other organic wastewater compounds in the influent water samples from wastewater-treatment plants in Miami-Dade County, Florida.

sulfonamide, had the greatest concentration range (0.029 to 1.7 µg/L) and highest median concentration (1.1 µg/L) while erythromycin-H$_2$O showed the smallest concentration range (0.009 to 0.22 µg/L), and tetracycline had the lowest median concentration (0.11 µg/L; table 4). Erythromycin was only detected in samples collected in the wet season (ranging from 0.008 to 0.01 µg/L) and sulfathiazole was detected once in the dry season (0.073 µg/L). The 17 antibiotics (enrofloxacin, lomefloxacin, sarafloxacin, roxithromycin, tylosin, virginiamycin, sulfachloropyridazine, sulfadimethoxine, sulfamethazine, chlorotetracyline, oxytetracycline, epi-chlorotetracycline, epi-iso-chlorotetracycline, epi-oxytetracycline, iso-chlorotetracycline, chloramphenicol, and ormetoprim) were not detected in any influent sample collected, although the poor recovery of chloramphenicol (0 percent) and virginiamycin (0 percent) in the influent spike sample (app. 1) affected the probability for their detection. The hormone E2 was detected in all but one influent sample (ranging from 0.011 to 0.22 µg/L, median of 0.042 µg/L), and was detected at higher concentrations in the dry season (table 4).

Pharmaceuticals and Other Organic Wastewater Compounds Detected in 24-Hour Flow-Weighted Effluent Composite Samples

Organic compounds in all classes were detected in effluent samples collected at each plant (table 5). Similar to the influent samples, the total concentrations in effluent samples collected in the dry season were generally 10 to 30 percent higher than samples collected in the wet season, with the exception of SD 1 (fig. 4A; app. 4). Estimated loads ranged from 0.3 to 25.7 lb/d and were higher in the wet season at four of the six plants (fig. 4B; app. 4). Wastewater-indicator compounds accounted for greater than 64 percent of the total concentration in each effluent sample while E2 accounted for less than 0.1 percent of the total concentration (app. 4). For graphical purposes, E2 was not included in figure 4.

Nineteen SVOCs were detected in at least one effluent sample (table 5). The compound 1,4-dichlorobenzene was detected in every 24-hr flow-weighted effluent sample (ranging from 0.12 to 2.3 µg/L). Concentrations of 1,4-dichlorobenzene in effluent from the WWTPs were well below the U.S. Environmental Protection Agency (USEPA) maximum contaminant level (MCL) of 75 µg/L for drinking water. Trace amounts (ranging from 0.021 to 0.14 µg/L) of 2,4,6-trichlorophenol were also detected in every effluent sample. The compound 2,4-dimethylphenol was detected in all but one effluent samples (ranging from 0.05 to 0.45 µg/L). The remaining 16 compounds were detected less frequently at low concentrations (less than or equal to 1 µg/L).

Thirteen pesticide compounds were detected in at least one effluent sample (table 5). The three pesticide compounds 3,4-dichloroaniline, fipronil, and fipronil sulfide, were detected in every effluent sample. The greatest concentration range

(0.046 to 0.24 µg/L) was for 3,4-dichloroaniline, a pesticide degradate, and highest median concentration (0.1 µg/L) while fipronil sulfide had the smallest concentration range (0.006 to 0.019 µg/L) and lowest median concentration (0.01 µg/L). Similar to the influent samples, carbaryl was detected at the highest concentrations (0.29 µg/L).

Forty-nine wastewater-indicator compounds were detected in at least one effluent sample (table 5). Caffeine showed the greatest concentration range (ranging from 0.053 to 25 µg/L) and the greatest concentration of all the target compounds in effluent samples. Caffeine is a stimulant in soft drinks, tea, coffee and energy drinks. A derivative of caffeine, 1,7-dimethylxanthine, was also detected as high as 1.3 µg/L. Tri(2-butoxyethyl) phosphate, a flame retardant, also showed a large concentration range (1.2 to 12 µg/L) and had the highest median concentration (3.1 µg/L).

Among the pharmaceuticals, seven were detected in effluent samples (table 5). Ibuprofen had the greatest concentration range (less than 0.05 to 1.3 µg/L) and highest maximum concentration. All other compounds were detected at concentrations less than or equal to 0.25 µg/L. Carbamazepine was the only pharmaceutical detected in every effluent sample collected (0.12 to 0.25 µg/L). The high detection of carbamazepine in effluent waters is consistent with previous studies that concluded carbamazepine to be neither subjected to degradation nor to adsorption processes during wastewater treatment (Clara and others, 2004).

Eleven antibiotics were detected in at least one effluent sample (table 5). Trimethoprim, azithromycin, ciprofloxacin, and ofloxacin were detected in every effluent sample collected (0.03 to 1.2 µg/L). Sulfamethoxazole (a sulfonamide) had the greatest concentration range (less than 0.005 to 1.9 µg/L). Doxycycline, sulfadiazine, and tylosin were only detected in effluent samples collected in the wet season. The hormone E2 was detected in 10 of the 12 effluent samples (ranging in concentration from 0.0097 to 0.08 µg/L).

Percent Reduction in Concentrations

One purpose of this report is to assess the percent reduction in concentration of pharmaceuticals and other organic wastewater compounds during treatment at each plant. The percent reduction (PR) is defined as:

$$PR = \frac{(C_I - C_E)}{C_I} \cdot 100 \qquad (2)$$

where C_I is the concentration of a compound in the influent, and C_E is the concentration of a compound in the effluent. There are a number of physical, chemical, and biological processes (for example, biodegradation, volatilization, photolysis, and adsorption, and so forth) that occur during wastewater treatment that can contribute to the overall percent reduction, as well as a number of factors (for example, temperature, pH,

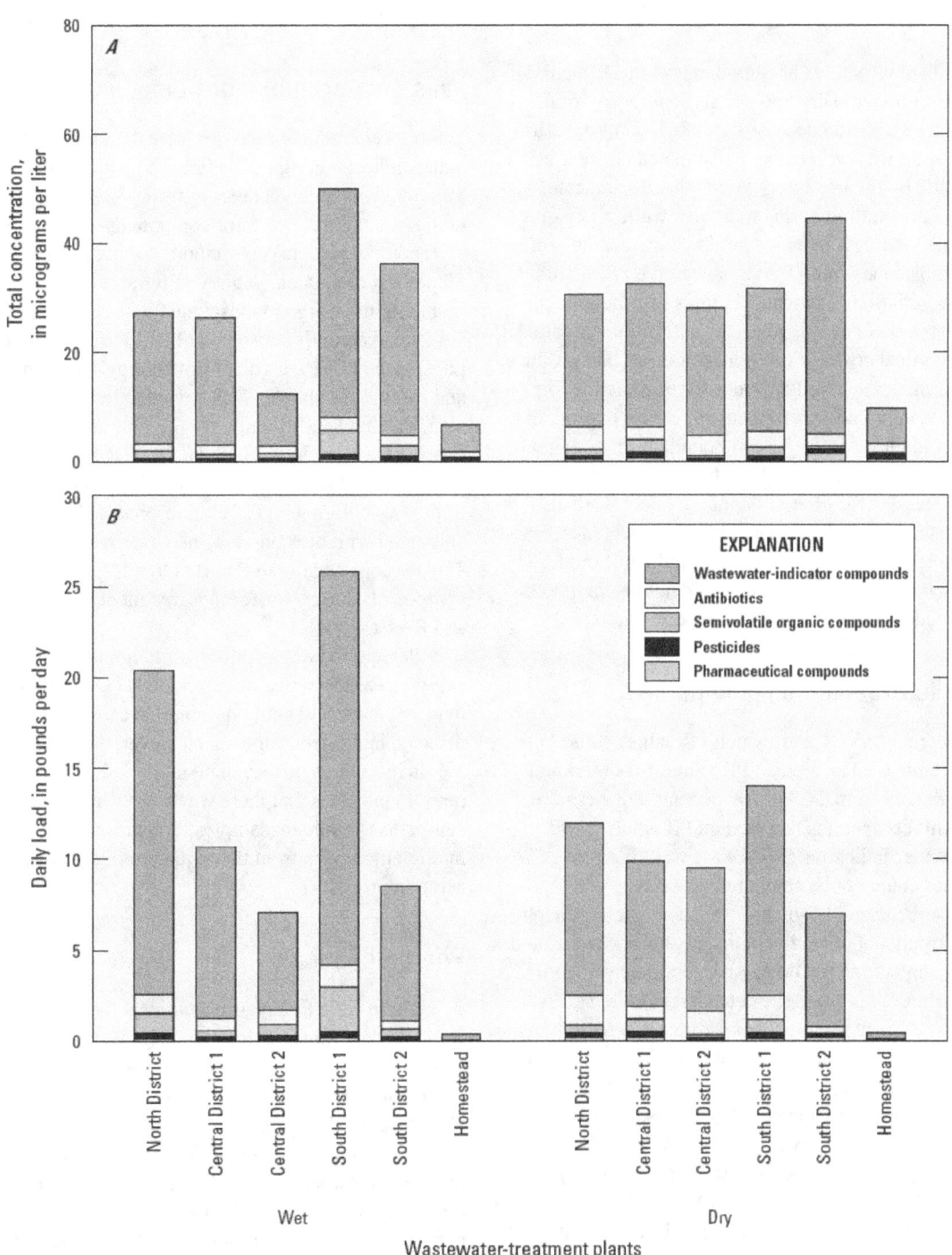

Figure 4. Total concentration *A*, and daily loads *B*, of pharmaceuticals and other organic wastewater compounds in the effluent water samples from wastewater-treatment plants in Miami-Dade County, Florida.

solids retention time, hydraulic retention time, and biomass concentration) that can affect the rates of these processes. This section summarizes the percent reduction in concentration of pharmaceuticals and other organic wastewater compounds at the effluent at each plant in both the dry and wet seasons.

Table 6 lists PR values of compounds at the effluents of the six plants in both the dry and wet seasons. Many of the PR values were estimated because the influent and(or) effluent concentration was reported as an estimated concentration reported as "E" by the laboratory. A positive PR indicates a decrease in concentration during treatment while a negative PR indicates an increase in concentration during treatment. Several circumstances could lead to a negative PR, including: (1) large variations in chemical inputs that the influent sampling design was not able to capture; (2) minor absolute sampling/analytical errors at low concentrations that produced relatively strongly negative PR values for recalcitrant compounds; (3) a compound was formed, either as a byproduct or from the degradation of a larger compound, during treatment processes; and (4) changes in solubility conditions during the treatment processes (for filtered samples). There were also numerous instances where a compound was detected in the effluent sample and not in the associated influent sample. The percent reduction in these instances could not be computed due to censored influent data and was reported as "C."

Semivolatile Organic Compounds

There were 20 SVOCs for which PR values could be determined (table 6). The median PR values for individual compounds ranged from 24.8 to 100 percent. Furthermore, 15 (75 percent) compounds had median PR values of 90 percent or greater indicating SVOCs are well removed. The PR values could not be computed for seven SVOCs (1,2,4-trichlorobenzene, 1,3-dichlorobenzene, 2-chloronaphthalene, 2-nitrophenol, dimethyl phthalate, nitrobenzene, and n-nitrosodimethylamine (NDMA)), because they were only detected in effluent samples. The detection of these seven compounds in effluent samples, along with nondetection in the complementary influent sample, could be an indication that these compounds are byproducts of the treatment processes or are degradation products of larger compounds. For example, studies have shown that NDMA is formed when wastewater is disinfected with chlorine, especially when ammonia is present (Pehlivanoglu-Mantas and others, 2006). NDMA was only present in the effluent at the three plants (CD1, CD2, and ND) that use chlorination to disinfect the treated wastewater.

Pesticides and Pesticide Degradates

There were 12 pesticides and pesticide degradates for which PR values could be determined (table 6). Percent reduction values were highly variable for pesticides and pesticide

degradates with median values for individual compounds ranging from -16.1 to 100 percent. Percent reduction values could not be calculated for one pesticide, metribuzin, and one pesticide degradate, 2-chloro-4-isopropylamino-6-amino-s-triazine (CIAT), because they were only detected in effluent samples.

Wastewater-Indicator Compounds

Percent reduction values were determined for 52 wastewater-indicator compounds (table 6) with median values for individual compounds ranging from -32.3 to 100 percent. Generally, wastewater-indicator compounds were removed with 36 (about 70 percent) compounds having median PR values of 90 percent or greater. Negative PR values for 11 compounds, including the 4 organophosphate flame retardant compounds (tri(2-butoxyethyl) phosphate (TBEP), tri(2-chloroethyl) phosphate (TCEP), tri(dichloroisopropyl) phosphate (TDCP), and tributyl phosphate (TBP), were noted on at least one occasion. The median PR value for the four organophosphate flame retardant compounds TBEP, TCEP, TDCP, and TBP were 55.9, -32.4, -24.9, and 14.7, respectively. The lower PR values of the two chlorinated organophosphates (TCEP and TDCP) compared with the non-chlorinated derivatives (TBEP and TBP) is consistent with reported trends in removal for these compounds during wastewater treatment in Germany (Meyer and Bester, 2004).

Percent reduction values could not be calculated for two wastewater-indicator compounds (3,4-dichlorophenyl isocyanate and anthraquinone) because they were only detected in effluent samples. However, the nondetection of 3,4-dichlorophenyl isocyanate in the influent spike sample (app. 1) indicates that there was a problem detecting this compound in influent samples. The presence of anthraquinone in effluent may be from the oxidation of anthracene in the aeration process.

Pharmaceuticals

There were five pharmaceuticals for which PR values could be determined at least once (table 6) with median values for individual compounds ranging from -24.5 to 100 percent. Percent reduction values were highly variable for pharmaceuticals. For example, acetaminophen (99.8 to 100 percent) was effectively removed during treatment at each plant while carbamazepine (-321 to 19.9 percent) and codeine (-1,120 to -66.0 percent) had very low and highly variable removal rates. Percent reduction values could not be calculated for three pharmaceuticals (dehydronifedipine, diltiazem, and diphenhydramine) because they were only detected in effluent samples. There were also four occasions when PR values could not be calculated for codeine. In a similar treatment study using the same analytical method, Lietz and Meyer (2006) reported higher concentrations of codeine, diltiazem, and diphenhydramine in effluent samples compared

Table 6.   Percent reduction in concentrations between influent and effluent samples collected from wastewater-treatment plants.

[ND, North District Wastewater Treatment Plant; CD1, Central District Wastewater Treatment Plant 1; CD2, Central District Wastewater Treatment Plant 2; SD1, South District Wastewater Treatment Plant 1; SD2, South District Wastewater Treatment Plant 2; HS, Homestead Wastewater Treatment Plant 2. Values are in percent. Negative percent reduction in red. Compounds in bold were only detected in effluent samples. n, number of samples; Mgal/d, million gallons per day; E, estimated due to one or more estimate data value; C, unable to compute percent reduction due to censored influent data; NWQL, U.S. Geological Survey National Water Quality Laboratory; CLLE, Continuous liquid-liquid extraction; GC/MS, gas chromatography/mass spectrometry; OGRL, U.S. Geological Survey Organic Geochemistry Research Laboratory; SPE, solid-phase extraction; HPLC/MS high-performance liquid chromatogrpahy/mass spectrometry; ELISA, enzyme-linked immunosorbent assay; —, not applicable]

Plant	ND		CD1		CD2		SD1		SD2		HS		Summary statistics			
Season	Wet	Dry	Wet	Dry	Wet	Dry	Wet	Dry	Wet	Dry	Wet	Dry	n	Minumum	Median	Maximum
Flow (Mgal/d)	90.4	47.0	48.8	36.3	69.7	40.6	62.1	53.2	28.2	18.8	5.6	4.5	12	4.5	43.8	90.4
Semivolatile organic compounds (NWQL, Unfiltered, CLLE, GC/MS)																
1,2,4-Trichlorobenzene	—	—	E C	—	E C	—	—	—	—	—	—	E C	0	—	—	—
1,2-Dichlorobenzene	E C	—	E C	—	E C	E 24.0	—	25.6	—	E 100	E C	E C	2	24	24.8	25.6
1,3-Dichlorobenzene	E C	—	E C	—	E C	—	—	—	—	—	—	E C	0	—	—	—
1,4-Dichlorobenzene	69.6	E 79.5	E 58.4	92.8	E 60.4	E 76.5	E -177.0	E 25.7	E 47.2	E 66.9	E 90.5	E 91.2	12	-177	68.3	92.8
2,4,6-Trichlorophenol	E 26.3	E C	E 45.4	E C	E 1.7	E C	E C	E 24.5	E C	E 71.3	E 79.2	E 76.1	7	1.7	45.4	79.2
2,4-Dichlorophenol	E 77.1	E 87.4	E 82.7	E 87.7	E 17.3	100	E -26.1	E 76.5	E 73.1	E 88.7	E 95.2	E 87.9	12	-26.1	85.1	100
2,4-Dimethylphenol	—	E C	E 91.8	E C	E 86.7	E 100	E C	E C	—	E C	E C	—	3	86.7	91.8	100
2-Chloronaphthalene	E C	—	—	—	—	—	—	—	—	—	—	—	0	—	—	—
2-Chlorophenol	—	E 100	—	—	—	—	—	—	—	—	—	—	1	100	100	100
2-Nitrophenol	—	—	—	—	—	—	E C	—	C	—	—	—	0	—	—	—
4-Chloro-3-methyl-phenol	—	—	100	E 99.6	100	E 100	—	—	100	—	—	—	3	100	100	100
Acenaphthene	E 100	E 100	E 100	—	E 100	E 100	E 100	E 100	100	E 100	E 89.0	E 95.8	11	89	100	100
Benz[a]anthracene	E 100	—	E 100	—	E 100	E 100	—	—	—	E 100	—	—	4	100	100	100
Benzo[b]fluoranthene	E 100	—	—	—	—	—	—	—	—	—	—	E 100	2	100	100	100
Benzo[k]fluoranthene	—	—	—	—	—	—	—	—	—	—	—	E 100	1	100	100	100
Butylbenzyl phthalate	E 100	—	E 100	—	E 100	—	E 64.1	E 100	E 100	—	E 100	E 100	8	64.1	100	100
Chrysene	E 100	—	E 100	E 100	E 100	E 100	—	E 100	—	E 100	—	E 100	7	100	100	100
Diethyl phthalate	100	100	100	E 100	100	100	E 100	100	100	100	100	100	11	100	100	100
Diethylhexyl phthalate (DEHP)	E 100	—	100	E 99.6	E 100	E 100	E 86.8	E 100	E 91.8	E 100	100	E 100	11	86.8	100	100
Dimethyl phthalate	E C	—	—	—	E C	—	—	E C	—	—	—	—	0	—	—	—
Di-n-butyl phthalate	E 100	—	E 100	—	E 100	—	E 71.4	—	E 100	—	—	E 100	6	71.4	100	100
Di-n-octyl phthalate	E 100	—	—	—	E 100	E 100	E 65.4	—	E 100	—	E 100	E 100	7	65.4	100	100

Table 6.   Percent reduction in concentrations between influent and effluent samples collected from wastewater-treatment plants.—Continued

[ND, North District Wastewater Treatment Plant; CD1, Central District Wastewater Treatment Plant 1; CD2, Central District Wastewater Treatment Plant 2; SD1, South District Wastewater Treatment Plant 1; SD2, South District Wastewater Treatment Plant 2; HS, Homestead Wastewater Treatment Plant. Values are in percent. Negative percent reduction in red. Compounds in bold were only detected in effluent samples. n, number of samples; Mgal/d, million gallons per day; E, estimated due to one or more estimate data value; C, unable to compute percent reduction due to censored influent data; NWQL, U.S. Geological Survey National Water Quality Laboratory; CLLE, Continuous liquid-liquid extraction; GC/MS, gas chromatogrpahy/mass spectrometry; OGRL, U.S. Geological Survey Organic Geochemistry Research Laboratory; SPE, solid-phase extraction; HPLC/MS high-performance liquid chromatogrpahy/mass spectrometry; ELISA, enzyme-linked immunosorbent assay; —, not applicable]

Plant	ND		CD1		CD2		SD1		SD2		HS		Summary statistics			
Season	Wet	Dry	Wet	Dry	Wet	Dry	Wet	Dry	Wet	Dry	Wet	Dry	n	Minumum	Median	Maximum
Fluorene	—	E 90.5	E 93.2	—	E 88.4	E 92.5	E 100	—	E 100	—	E 93.6	E 95.2	8	88.4	93.4	100
Nitrobenzene	E C	—	—	—	—	—	—	E C	—	—	E C	—	0	—	—	—
N-Nitrosodimethyl-amine (NDMA)	—	E C	E C	—	E C	—	—	E C	—	—	E C	—	0	—	—	—
N-Nitrosodiphenyl-amine	—	—	—	—	—	E 100.0	—	—	—	—	—	—	1	100	100	100
Pentachlorophenol	E 100.0	—	—	—	—	—	—	—	—	—	—	E 42.0	2	42	71	100
Pesticides and pesticide degradates (NWQL, Filtered, SPE, GC/MS)																
1-Naphthol	E 100	E C	—	—	—	—	—	E 76.0	—	E 100	—	—	3	76	100	100
2-Chloro-4-isopropyl-amino-6-amino-s-triazine (CIAT)	—	—	—	—	—	—	—	—	—	—	E C	E C	0	—	—	—
3,4-Dichloroaniline	E C	E C	E C	E C	E 18.6	E C	E C	E C	E-131	E-5.0	E-80.4	E C	4	-131	-5	34.1
Atrazine	C	C	C	C	E C	—	-0.8	—	2	—	4.6	E 20.0	4	-0.8	3.3	20
Carbaryl	E 80.2	E C	E 14.9	E-14.8	E 15.0	E C	E C	E C	E C	E C	E 88.5	E C	5	-14.8	15	88.5
Chlorpyrifos	—	—	—	—	—	—	E 56.1	—	E 53.6	E C	E 24.1	—	3	24.1	53.6	56.1
Desulfinylfipronil	E 100	E C	E C	E 16.7	E C	E C	E-5.4	E 12.5	E 8.3	E 16.7	E-71.4	E-117	8	-117	10.4	100
Fipronil	E-5.6	E-8.7	E 2.4	E 24.8	E-12.2	E-1.2	E 1.9	E 6.8	E-8.8	E-32.6	E 68.7	E-12.5	12	-32.6	-3.4	68.7
Fipronil sulfide	E 0.0	E C	E-25.0	-35.7	E-25.0	E-37.5	E-18.0	E 0.0	E-16.1	E 11.1	E 17.5	E 0.0	11	-37.5	-16.1	17.5
Fipronil sulfone	—	—	—	E C	E C	—	—	E C	E 100	E C	—	—	1	100	100	100
Hexazinone	—	—	17.6	—	26.1	—	-13.3	-9.1	-9.5	-7.1	—	—	6	-13.3	-8.1	26.1
Methidathion	—	—	—	—	—	—	—	—	—	—	100	—	1	100	100	100
Metribuzin	—	—	—	—	—	—	—	—	—	C	—	—	0	—	—	—
Terbuthylazine	—	—	—	—	—	—	—	—	—	—	50.7	—	1	50.7	50.7	50.7
Wastewater-indicator compounds (NWQL, Unfiltered, CLLE, GC/MS)																
1,7-Dimethylxanthine	E 100	100	100	E 90.2	100	88.3	E 91.4	100	E 100	85.6	E 100	98.3	12	85.6	85.6	100
1-Methylnaphthalene	E 100	100	E 100	—	100	E 97.9	E 100	E 100	E 100	E 100	E 100	E 100	11	97.9	97.9	100

Table 6. Percent reduction in concentrations between influent and effluent samples collected from wastewater-treatment plants.—Continued

[ND, North District Wastewater Treatment Plant; CD1, Central District Wastewater Treatment Plant 1; CD2, Central District Wastewater Treatment Plant 2; SD1, South District Wastewater Treatment Plant 1; SD2, South District Wastewater Treatment Plant 2; HS, Homestead Wastewater Treatment Plant. Values are in percent. Negative percent reduction in red. Compounds in bold were only detected in effluent samples. n, number of samples; Mgal/d, million gallons per day; E, estimated due to one or more estimate data value; C, unable to compute percent reduction due to censored influent data; NWQL, U.S. Geological Survey National Water Quality Laboratory; CLLE, Continuous liquid-liquid extraction; GC/MS, gas chromatography/mass spectrometry; OGRL, U.S. Geological Survey Organic Geochemistry Research Laboratory; SPE, solid-phase extraction; HPLC/MS high-performance liquid chromatography/mass spectrometry; ELISA, enzyme-linked immunosorbent assay; —, not applicable]

Plant	ND		CD1		CD2		SD1		SD2		HS		Summary statistics			
Season	Wet	Dry	Wet	Dry	Wet	Dry	Wet	Dry	Wet	Dry	Wet	Dry	n	Minumum	Median	Maximum
2,6-Dimethylnaphthalene	E 100	100	E 97.2	—	E 98.1	E 100	E 100	E 100	E 100	E 100	E 100	E 100	11	97.2	100	100
2-Methylnaphthalene	100	100	E 100	—	100	E 98.2	E 100	E 100	E 100	100	E 100	E 100	11	98.2	100	100
3,4-Dichlorophenyl isocyanate	EC	EC	EC	—	EC	EC	EC	EC	EC	EC	EC	EC	0	—	—	—
3-*beta*-Coprostanol	E 95.9	E 97.4	E 100	—	E 100	E 99.3	E 96.2	E 97.2	E 96.6	E 93.6	E 100	E 99.6	11	93.6	97.4	100
3-Methyl-1h-indole (skatol)	E 99.0	100	E 100	—	100	E 93.1	E 94.3	E 100	E 96.1	E 99.0	E 99.7	E 99.1	11	93.1	99.1	100
3-*tert*-Butyl-4-hydroxyanisole (BHA)	—	E 56.2	E 100	—	E 100	E -54.0	E -78.8	E 36.6	E 100	1.9	E 100	E 100	10	-78.8	78.1	100
4-Cumylphenol	—	E 100	—	—	—	—	—	—	E 100	—	E 100	—	3	100	100	100
4-Nonylphenol (total)	E 97.8	E 98.1	E 94.7	—	E 100	E 93.5	E 90.0	E 94.5	E 89.0	E 87.1	E 87.5	E 100	11	87.1	94.5	100
4-Nonylphenol diethoxylate (NP$_2$EO)	EC	E 90.6	E 19.3	—	E 100	E -21.7	EC	EC	EC	EC	E 100	EC	5	-21.7	90.6	100
4-Nonylphenol monoethoxylate (NP$_1$EO)	E 93.5	E 77.5	—	—	E 100	E 92.1	E 86.8	EC	EC	EC	—	E 100	6	77.5	92.8	100
4-Octylphenol diethoxylate (OP$_2$EO)	E 77.8	E 96.2	E 59.9	—	E 83.8	E 86.3	E 74.9	E 97.0	E 84.4	98.3	E 99.2	100	11	59.9	86.3	100
4-Octylphenol monoethoxylate (OP$_1$EO)	E 83.5	E 100	E 78.7	—	E 79.1	E 52.1	EC	E 100	EC	E 100	E 100	—	8	52.1	91.8	100
4-*tert*-octylphenol	E 97.4	E 100	E 96.6	—	E 100	E 100	E 92.6	E 100	E 93.4	E 100	E 96.4	—	10	92.6	98.7	100
5-methyl-1h-benzotriazole	E 80.4	—	EC	—	EC	E -73.9	EC	EC	EC	EC	—	—	2	-73.9	3.2	80.4
Acetophenone	E 80.7	100	E 100	—	100	100	E -19.9	E 100	E 40.6	100	E 41.3	—	10	-19.9	3.2	100
Anthracene	EC	E 100	E 100	—	E 100	—	—	E 100	—	E 84.4	E 100	E 100	7	84.4	100	100
Anthraquinone	EC	—	—	—	—	—	—	—	—	—	EC	EC	0	—	—	—

Table 6. Percent reduction in concentrations between influent and effluent samples collected from wastewater-treatment plants.—Continued

[ND, North District Wastewater Treatment Plant; CD1, Central District Wastewater Treatment Plant 1; CD2, Central District Wastewater Treatment Plant 2; SD1, South District Wastewater Treatment Plant 1; SD2, South District Wastewater Treatment Plant 2; HS, Homestead Wastewater Treatment Plant Values are in percent. Negative percent reduction in red. Compounds in bold were only detected in effluent samples. n, number of samples; Mgal/d, million gallons per day; E, estimated due to one or more estimate data value; C, unable to compute percent reduction due to censored influent data; NWQL, U.S. Geological Survey National Water Quality Laboratory; CLLE, Continuous liquid-liquid extraction; GC/MS, gas chromatograpy/mass spectrometry; OGRL, U.S. Geological Survey Organic Geochemistry Research Laboratory; SPE, solid-phase extraction; HPLC/MS high-performance liquid chromatograpy/mass spectrometry; ELISA, enzyme-linked immunosorbent assay; —, not applicable]

Plant	ND		CD1		CD2		SD1		SD2		HS		Summary statistics			
Season	Wet	Dry	Wet	Dry	Wet	Dry	Wet	Dry	Wet	Dry	Wet	Dry	n	Minumum	Median	Maximum
Benz[a]pyrene	E 87.5	—	E 100	—	E 100	E 100	E 100	—	E 100	—	—	—	6	87.5	100	100
Benzophenone	77.5	73.8	E 65.2	—	E 65.4	E 64.4	E 1.8	E 58.9	E 37.3	E 52.6	E 79.6	E 100	11	1.8	65.2	100
beta-Sitosterol	E 91.9	E 92.6	E 100	—	E 100	E 94.3	E 90.4	E 94.2	E 83.9	E 86.5	E 100	E 93.9	11	83.9	93.9	100
beta-Stigmastanol	E 92.7	E 100	E 100	—	E 100	E 100	E 100	E 100	E 100	E 100	E 100	E 100	11	92.7	100	100
Bisphenol A	E 81.5	E C	E 88.9	—	E 96.7	E C	E 93.6	C	E 97.0	E 98.2	100	100	8	81.5	97	100
Bromoform	E C	E C	E C	—	E C	—	—	—	—	—	E 78.0	E C	1	78.0	78.0	78.0
Caffeine	E 98.9	97.5	E 98.8	E 40.0	E 97.1	E 98.2	E 74.7	97.6	E 98.5	91.0	E 99.9	E 99.7	12	40.0	97.9	99.9
Camphor	E 99.2	100	E 100	—	100	E 100	E 98.2	E 100	E 100	E 96.3	E 100	E 100	11	96.3	100	100
Carbazole	—	—	—	—	—	—	—	—	—	—	E 100	E 100	2	100	100	100
Cholesterol	E 97.8	E 97.5	E 94.8	—	E 100	E 99.2	E 97.0	E 97.0	E 95.5	E 94.2	E 100	E 99.3	11	94.2	97.5	100
Cotinine	—	E C	—	—	E C	E C	E 60.5	—	E 86.1	E C	E 100	E 100	4	60.5	93.0	100
Fluoranthene	E 91.4	E 100	E 90.5	—	E 100	E C	E 69.8	E C	E 62.7	E C	E 100	E 100	8	62.7	95.7	100
Galaxolide (HHCB)	74.5	58.6	E 60.2	—	E 61.0	E 51.8	E 37.4	E 52.1	E -1.9	E 43.6	E 72.9	E 39.4	11	-1.9	52.1	74.5
Indole	E 100	100	E 100	—	E 100	E 100	E 93.3	—	E 98.4	—	E 100	E 100	9	93.3	100	100
Isoborneol	100	100	E 100	—	100	E 100	E 100	E 100	E 100	100	E 100	E 100	11	100	100	100
Isophorone	E C	—	E 94.1	—	E C	E C	E C	E C	E C	—	E C	—	1	94.1	94.1	94.1
Isopropylbenzene (cumene)	E 100	E 100	E 100	—	E 100	E 100	—	—	—	—	E 100	—	6	100	100	100
Limonene	E 100	E 100	E 100	—	E 100	E 99.1	E 100	E 100	E 100	E 100	E 100	E 100	11	99.1	100	100
Menthol	E 100	100	E 98.1	—	E 98.8	E 100	E 97.4	E 100	E 100	100	E 99.6	E 100	11	97.4	100	100
Methyl salicylate	100	—	E 100	—	100	—	E 100	E 100	E 100	100	E 96.6	E 100	9	96.6	100	100
N,N-Diethyl-meta-toluamide (DEET)	—	100	E 63.8	—	E 89.1	E 42.5	E 56.1	E 86.6	E 88.0	E 70.8	E 97.4	E 100	10	42.5	87.3	100
Naphthalene	100	100	E 100	—	100	E 95.6	E 100	E 100	E 100	100	E 94.8	E 100	11	94.8	100	100
para-Cresol	E 99.1	E 99.6	E 90.9	—	E 99.6	E 95.1	E 100	—	E 99.9	100	E 100	—	9	90.9	99.6	100
Phenanthrene	100	E 100	E 100	—	E 100	E 100	E 88.7	E 100	E 91.6	E 93.3	E 100	E 100	11	88.7	100	100
Phenol	100	E 100	—	—	100	—	—	—	E 100	100	—	—	10	100	100	100
Pyrene	E C	E C	E C	—	E C	E C	E 17.6	E C	E C	E C	E 100	—	2	17.6	58.8	100
Tetrachloroethylene	E 53.3	E 58.2	E 36.7	—	E 64.1	E 17.6	E 20.5	E 7.8	E 14.0	E -16.4	E 100	E 100	11	-16.4	36.7	100

Table 6. Percent reduction in concentrations between influent and effluent samples collected from wastewater-treatment plants.—Continued

[ND, North District Wastewater Treatment Plant; CD1, Central District Wastewater Treatment Plant 1; CD2, Central District Wastewater Treatment Plant 2; SD1, South District Wastewater Treatment Plant 1; SD2, South District Wastewater Treatment Plant 2; HS, Homestead Wastewater Treatment Plant. Values are in percent. Negative percent reduction in red. Compounds in bold were only detected in effluent samples. n, number of samples; Mgal/d, million gallons per day; E, estimated due to one or more estimate data value; C, unable to compute percent reduction due to censored influent data; NWQL, U.S. Geological Survey National Water Quality Laboratory; CLLE, Continuious liquid-liquid extraction; GC/MS, gas chromatogrpahy/mass spectrometry; OGRL, U.S. Geological Survey Organic Geochemistry Research Laboratory; SPE, solid-phase extraction; HPLC/MS high-performance liquid chromatography/mass spectrometry; ELISA, enzyme-linked immunosorbent assay; —, not applicable]

Plant	ND		CD1		CD2		SD1		SD2		HS		n	Summary statistics		
Season	Wet	Dry	Wet	Dry	Wet	Dry	Wet	Dry	Wet	Dry	Wet	Dry		Minumum	Median	Maximum
Tonalide (AHTN)	E C	E C	E C	—	E 87.8	E 84.5	E 62.5	E 66.6	E 50.8	E C	E 85.9	E 100	7	50.8	84.5	100
Tri(2-butoxyethyl) phosphate (TBEP)	E 54.9	E 55.9	E 19.9	—	E 57.3	E 62.8	E -15.7	E 57.8	E -23.2	E 35.0	E 91.2	E 72.7	11	-23.2	55.9	91.2
Tri(2-chloroethyl) phosphate (TCEP)	E C	—	E C	—	E 41.3	E -70.3	E -38.8	—	E -30.5	0.0	E -34.2	—	5	-70.3	-32.3	41.3
Tri(dichloroisopropyl) phosphate (TPIP)	-18.9	19.2	E -21.7	—	E 24.9	E -55.1	E -57.1	E -47.0	E -47.2	-28.2	E 1.2	—	10	-57.1	-24.9	24.9
Tributyl phosphate (TBP)	—	E C	E C	—	E 64.8	E 14.7	E -29.7	E C	E -41.5	E C	E 39.8	—	5	-41.5	14.7	64.8
Triclosan	93.9	82.1	E 100	—	100	E 88.3	E 77.8	E 100	E 84.4	E 87.1	E 100	E 100	11	77.8	93.9	100
Triethyl citrate (ethyl citrate)	61.5	50.6	E 22.0	—	E 41.6	E -64.9	E -21.0	36.0	E 15.4	E 32.0	E 71.1	E 100	11	-64.9	36.0	100
Triphenyl phosphate	E 68.4	E 36.9	E 36.3	—	E 44.6	E 25.4	E 0.0	23.8	E 0.0	E 1.3	E 76.9	—	10	0.0	30.9	76.9
Pharmaceutical compounds (NWQL, Filtered, SPE, HPLC/MS)																
Acetaminophen	E 99.8	100	E 100	E 100	E 100	E 100	E 100	100	E 100	100	E 100	E 100	12	99.8	100	100
Carbamazepine	10.3	-40.4	-321	-26.1	-24.5	-56.0	-21.5	-35.2	-17.0	-3.6	19.9	—	11	-321	-24.5	19.9
Codeine	—	E C	E C	E C	E C	—	E -1,120	E -70.4	E C	E -96.4	E -66.0	E -389	5	-1,120	-96.4	-66.0
Dehydronifedipine	—	E C	—	E C	—	E C	—	E C	E C	E C	E C	E C	0	—	—	—
Diltiazem	—	E C	—	E C	E C	E C	E C	E C	E C	E C	—	E C	0	—	—	—
Diphenhydramine	—	E C	—	—	E C	—	—	—	—	—	—	—	0	—	—	—
Ibuprofen	100	-166	100	88.1	100	100	—	100	—	76.9	—	95.4	9	-166	100	100
Thiabendazole	—	—	—	—	—	—	—	—	—	—	E 100	—	1	100	100	100
Antibiotics (OGRL, Filtered, SPE, LC/MS)																
Azithromycin	C	-987	-24.9	-117	—	-275	30.1	C	65.7	C	66.8	47.9	9	-987	-24.9	66.8
Ciprofloxacin	45.9	-220	76.8	21.0	89.4	-145	26.5	52.1	53.9	-16.9	95.0	86.9	12	-220	49.0	95.0
Doxycycline	—	—	100	—	100	100	C	—	C	—	C	100	4	100	100	100
Epi-tetracycline	—	—	100	—	—	—	—	—	—	—	—	100	2	100	100	100
Erythromycin	—	—	—	—	C	—	-111	—	-60.0	—	-150	—	3	-150	-111	-60.0
Erythromycin-H₂O	72.4	-10.8	-350	32.7	46.9	26.7	-425	47.9	-88.5	100	-111	53.7	12	-425	29.7	100

Table 6. Percent reduction in concentrations between influent and effluent samples collected from wastewater-treatment plants.—Continued

[ND, North District Wastewater Treatment Plant; CD1, Central District Wastewater Treatment Plant 1; CD2, Central District Wastewater Treatment Plant 2; SD1, South District Wastewater Treatment Plant 1; SD2, South District Wastewater Treatment Plant 2; HS, Homestead Wastewater Treatment Plant. Values are in percent. Negative percent reduction in red. Compounds in bold were only detected in effluent samples. n, number of samples; Mgal/d, million gallons per day; E, estimated due to one or more estimate data value; C, unable to compute percent reduction due to censored influent data; NWQL, U.S. Geological Survey National Water Quality Laboratory; CLLE, Continuous liquid-liquid extraction; GC/MS, gas chromatograpy/mass spectrometry; OGRL, U.S. Geological Survey Organic Geochemistry Research Laboratory; SPE, solid-phase extraction; HPLC/MS high-performance liquid chromatogrpahy/mass spectrometry; ELISA, enzyme-linked immunosorbent assay; —, not applicable]

Plant	ND		CD1		CD2		SD1		SD2		HS		Summary statistics			
Season	Wet	Dry	Wet	Dry	Wet	Dry	Wet	Dry	Wet	Dry	Wet	Dry	n	Minumum	Median	Maximum
Lincomycin	100	—	—	100	—	—	—	—	—	—	—	—	2	100	100	100
Norfloxacin	—	—	—	—	100	100	—	—	—	—	—	—	2	100	100	100
Ofloxacin	8.2	-114	-138	53.9	-10.8	-148	-1.1	31.7	32.0	31.5	50.3	30.1	12	-148	19.2	53.9
Sulfadiazine	—	—	—	—	—	—	69.5	—	—	—	65.1	100	3	65.1	69.5	100
Sulfamethoxazole	98.4	-399	3.3	-193	95.2	41.0	62.4	-318	100	86.3	65.4	99.6	12	-399	63.9	100
Sulfathiazole	—	—	—	—	—	—	—	100	—	—	—	—	1	100	100	100
Tetracycline	68.6	100	76.9	19.7	93.6	52.8	4.0	85.3	74.0	48.3	100	95.8	12	4.0	75.5	100
Trimethoprim	8.5	-221	-270	43.1	23.6	-279	33.7	12.2	14.3	34.1	95.6	78.4	12	-279	19.0	95.6
Tylosin	C	—	C	—	C	—	—	—	—	—	—	—	0	—	—	—
Hormone (OGRL, Filtered, ELISA)																
17-*beta*-Estradiol (E2)	67.6	-36.4	48.6	12.5	-325	—	73.7	95.6	-90.5	100	47.6	69.2	11	-325	48.6	100
Total																
Minimum	-18.9	-987	-350	-193	-596	-279	-1120	-318	-131	-96.4	-150	-389	—	—	—	—
Median	97.8	97.8	95.7	40.0	100	93.3	67.5	96.3	88.5	89.8	97.4	100	—	—	—	—
Maximum	100	100	100	100	100	100	100	100	100	100	100	100	—	—	—	—
Median values for each class																
Semivolatile organic compounds	100	95	100	100	100	100	71	100	100	100	95	100	—	—	—	—
Pesticides and pesticide degradates	80.2	-8.7	14.9	0.9	15.0	-19.4	-3.1	6.8	-3.4	3.1	24.1	-6.3	—	—	—	—
Wastewater-indicator compounds	97.4	100.0	98.8	65.1	100.0	93.5	89.4	100.0	95.8	94.2	100.0	100.0	—	—	—	—
Pharmaceutical compounds	99.8	-40.4	100.0	88.1	100.0	100.0	-21.5	32.4	41.5	36.6	59.9	95.4	—	—	—	—
Antibiotics	68.6	-220.1	3.3	26.8	89.4	26.7	26.5	47.9	43.0	41.2	65.4	91.3	—	—	—	—

to influent samples, and attributed the higher concentrations in effluent samples to solubility changes caused by matrix conditions and (or) issues with matrix interferences using the LC/MS method. Percent reduction values for ibuprofen were high (greater than 76.9 percent) during treatment for all but one sample occasion (NDWWTP, dry season).

Antibiotics

There were 14 antibiotics for which PR values could be determined at least once (table 6), with median values for individual compounds ranging from -111.1 to 100 percent. Similar to the pharmaceuticals, PR values were highly variable for antibiotics. For example, PR values for doxycycline were 100 percent on four occasions, but could not be calculated on three occasions due to nondetects in influent samples. Percent reduction values for sulfamethoxazole and erythromycin-H$_2$O were highly variable, ranging from -399 to 100 percent and -425 to 100 percent, respectively. Percent reduction values for four antibiotics (epi-tetracycline, lincomycin, norfloxacin, and sulfathiazole) were 100 percent when these compounds were detected in influent samples. Percent reduction values could not be calculated for one antibiotic (tylosin) because it was only detected in effluent samples.

17-*beta*-estradiol

Percent reduction values for 17-*beta*-estradiol (E2) were highly variable (ranging from -325 to 100 percent; table 6). This large variation in PR values is not consistent with other studies, which determined that E2 was effectively removed (greater than 98 percent) during treatment at activated sludge plants in Germany (Andersen and others, 2003). The reason for the large variation at the plants for this study is unknown and will require further investigation.

Pharmaceuticals and Other Organic Wastewater Compounds Detected in Groundwater, Canal Water, and Canal Bed Sediments

The HSWWTP has discharged treated effluent since the 1950s, initially to onsite percolation ponds, and more recently directly to the groundwater table using onsite rapid-rate soakage trenches. This section of the report summarizes the occurrence of pharmaceuticals and other organic wastewater compounds in groundwater, canal water, and bed sediments near the HSWWTP. The occurrence and concentrations of pharmaceuticals and other organic wastewater compounds in the dry season versus the wet season are compared.

Pharmaceuticals and Other Organic Wastewater Compounds in Groundwater near Effluent Soakage Trenches

Groundwater samples were collected from three monitoring wells (MW-1, MW-2, and MW-3; fig. 2) at variable distances from the soakage trenches to determine whether pharmaceuticals and other organic wastewater compounds that are not removed during treatment are being transported in the Biscayne aquifer. Samples were collected on two occasions, August 11, 2008 (wet season) and February 10-11, 2009 (dry season), and analyzed for the pharmaceuticals and other organic wastewater compounds listed in table 2.

In both the wet and dry seasons, groundwater from the three monitoring wells was suboxic with dissolved-oxygen (DO) concentrations ranging from 0.18 to 0.56 mg/L. Dissolved-oxygen concentrations in groundwater samples at this site exceeded the median DO concentration (0.14 mg/L) from 65 monitoring wells that tap the Biscayne aquifer (Lindsey and others, 2009).

Measureable concentrations of 41 compounds (including 6 SVOCs, 11 pesticides and pesticide degradates, 17 wastewater-indicator compounds, 3 pharmaceuticals, and 4 antibiotics) were detected in one or more groundwater samples (table 7; fig. 5). Fourteen compounds were detected in every groundwater sample collected (compounds shown in bold type, table 7). Generally, concentrations were below 0.5 μg/L with the exception of the fragrance HHCB (E 0.62–1.31 μg/L), the antibiotic sulfamethoxazole (0.14–0.57 μg/L), and the flame retardant tri(2-butoxyethyl)phosphate (E 0.51 μg/L) (table 7; fig. 5. The USEPA MCLs exist for a small set of these compounds, but none of the detected compounds exceeded MCLs for drinking water.

Generally, compounds detected in monitoring wells were detected at concentrations similar to or less than concentrations detected in the effluent at the HSWWTP (table 7; fig. 5); however, nine compounds (5-methyl-1h-benzotriazole, desulfinylfipronil amide, fipronil sulfone, hexachlorocyclopentadiene, metolachlor, prometon, simazine, sulfamethazine, and triclosan) were detected in groundwater samples that were not detected in corresponding effluent composite samples. This finding indicates that the source of these compounds is not the HSWWTP and(or) temporal variability in the removal efficiencies for these compounds during treatment at the HSWWTP. For example, 5-methyl-1H-benzotriazole and triclosan were frequently detected in effluent samples at the other plants as part of this study. Therefore, absence in effluent samples at the HSWWTP could mean that these compounds were not present at detectable levels during sample collection. The herbicides metolachlor, prometon, and simazine were not detected in any effluent composite sample for this study and their detection in groundwater at the HSWWTP site could be from the land application of these herbicides in nearby urban and agricultural areas. Several studies have reported the detection of metolachlor, prometon, and simazine in monitoring wells

Table 7. Pharmaceuticals and other organic wastewater compounds detected in the Homestead Wastewater Treatment Plant (HSWWTP) effluent samples and groundwater samples collected from three monitoring wells at the HSWWTP.

[Concentrations in micrograms per liter (μg/L). Compounds in bold indicates detection in every groundwater sample. HS, Homestead; EFF, effluent; MW, monitoring well; E, estimated value; NWQL, U.S. Geological Survey National Water Quality Laboratory; CLLE, Continuous liquid-liquid extraction; GC/MS, gas chromatogrpahy/mass spectrometry; OGRL, U.S. Geological Survey Organic Geochemistry Research Laboratory; SPE, solid-phase extraction; HPLC/MS high-performance liquid chromatogrpahy/mass spectrometry; ELISA, enzyme-linked immunosorbent assay; —, not detected]

Compound	HS EFF 20091020 1000 Wet	HS EFF 20090210 0200 Dry	MW-1 20080811 1500 Wet	MW-2 20080811 1100 Wet	MW-3 20080811 0900 Wet	MW-1 20090210 1530 Dry	MW-2 20090211 1500 Dry	MW-3 20090210 1300 Dry
Semivolatile organic compounds (NWQL, Unfiltered, CLLE, GC/MS)								
1,2-Dichlorobenzene	E 0.026	E 0.041	E 0.034	E 0.049	E 0.033	E 0.032	E 0.009	E 0.025
1,4-Dichlorobenzene	0.12	0.25	—	—	—	—	E 0.061	—
2,4,6-Trichlorophenol	E 0.021	E 0.086	—	—	—	E 0.037	—	—
2,4-Dichlorophenol	E 0.05	E 0.14	—	—	—	E 0.033	—	—
Hexachlorocyclopentadiene	—	—	—	—	—	E 0.03	—	E 0.024
Pentachlorophenol	—	E 0.23	—	—	—	E 0.22	—	E 0.24
Pesticides and pesticide Degradates (NWQL, Filtered, SPE, GC/MS)								
2-Chloro-4-isopropylamino-6-amino-s-triazine (CIAT)	E 0.018	E 0.021	E 0.021	E 0.020	E 0.021	E 0.020	E 0.017	E 0.022
3,4-Dichloroaniline (3,4-DCA)	E 0.095	E 0.17	E 0.11	E 0.2	E 0.15	E 0.086	E 0.14	E 0.092
Atrazine	0.019	0.032	E 0.032	E 0.033	E 0.034	E 0.011	E 0.015	E 0.023
Desulfinylfipronil	0.012	0.013	E 0.009	E 0.012	E 0.011	E 0.011	E 0.01	E 0.01
Desulfinylfipronil amide	—	—	E 0.008	E 0.008	E 0.008	E 0.002	E 0.002	E 0.003
Fipronil	E 0.016	E 0.027	E 0.014	E 0.019	E 0.014	E 0.015	E 0.008	E 0.007
Fipronil sulfide	E 0.007	E 0.006	E 0.008	E 0.008	E 0.011	E 0.006	E 0.010	E 0.011
Fipronil sulfone	—	—	E 0.007	E 0.008	E 0.008	E 0.006	E 0.006	E 0.006
Metolachlor	—	—	0.033	0.053	—	E 0.009	E 0.012	0.025
Prometon	—	—	—	—	E 0.008	E 0.009	E 0.009	E 0.011
Simazine	—	—	0.009	0.012	—	—	—	—
Wastewater-indicator compounds (NWQL, Unfiltered, CLLE, GC/MS)								
3-*beta*-Coprostanol	—	E 0.24	—	—	—	E 0.18	E 0.25	E 0.25
4-Nonylphenol (total, NP)	E 0.65	—	—	—	—	—	—	E 0.49
4-*tert*-Octylphenol (OP)	E 0.043	—	—	—	0.31	—	—	—
5-Methyl-1H-benzotriazole	—	—	—	—	—	E 0.15	E 0.24	E 0.23
Benzophenone	E 0.10	—	E 0.032	E 0.048	E 0.026	—	—	—
Bromoform	E 0.013	E 0.014	—	—	E 0.014	E 0.15	E 0.12	E 0.19

Table 7. Pharmaceuticals and other organic wastewater compounds detected in the Homestead Wastewater Treatment Plant (HSWWTP) effluent samples and groundwater samples collected from three monitoring wells at the HSWWTP.—Continued

[Concentrations in micrograms per liter (µg/L). Compounds in bold indicates detection in every groundwater sample. HS, Homestead; EFF, effluent; MW, monitoring well; E, estimated value; NWQL, U.S. Geological Survey National Water Quality Laboratory; CLLE, Continuous liquid-liquid extraction; GC/MS, gas chromatograpahy/mass spectrometry; OGRL, U.S. Geological Survey Organic Geochemistry Research Laboratory; SPE, solid-phase extraction; HPLC/MS high-performance liquid chromatogrpahy/mass spectrometry; ELISA, enzyme-linked immunosorbent assay; —, not detected]

Compound	Site name							
Sample date (yyyymmdd)	HS EFF 20091020	HS EFF 20090210	MW-1 20080811	MW-2 20080811	MW-3 20080811	MW-1 20090210	MW-2 20090211	MW-3 20090210
Sample Time	1000	0200	1500	1100	0900	1530	1500	1300
Season	Wet	Dry	Wet	Wet	Wet	Dry	Dry	Dry
Caffeine	E 0.024	E 0.071	—	—	—	E 0.078	—	—
Cholesterol	—	E 0.39	—	—	—	E 0.23	E 0.27	E 0.24
DEET	E 0.04	—	E 0.19	0.38	0.24	—	—	E 0.053
Galaxolide (HHCB)	0.8	E1.0	0.87	1.3	1.2	E 0.62	0.7	0.99
Methyl salicylate	E 0.02	—	—	E 0.017	—	—	—	—
Tonalide (AHTN)	E 0.045	—	E 0.066	E 0.10	E 0.12	—	—	E 0.11
Tri(2-butoxyethyl) phosphate (TBEP)	E1.3	E3.1	—	—	E 0.51	—	—	—
Tri(2-chloroethyl) phosphate	0.22	—	E 0.13	E 0.20	E 0.17	—	—	—
Tri(dichloroisopropyl) phosphate	0.34	—	E 0.20	0.33	0.31	—	—	—
Tributyl phosphate	E 0.097	—	E 0.023	E 0.033	E 0.029	—	—	—
Triclosan	—	—	E 0.059	E 0.071	—	—	—	—
Pharmaceutical compounds (NWQL, Filtered, SPE, HPLC/MS)								
Carbamazepine	0.19	0.2	0.12	0.15	0.17	0.15	0.13	0.13
Dehydronifedipine	E 0.006	E 0.007	—	—	—	E 0.006	E 0.004	—
Diphenhydramine	0.051	0.12	—	—	—	—	E 0.003	—
Antibiotics (OGRL, Filtered, SPE, LC/MS)								
Azithromycin	0.29	0.2	0.012	0.034	0.028	—	—	—
Ofloxacin	0.3	1.2	0.056	0.12	0.049	0.046	0.11	0.008
Sulfamethazine	—	—	—	—	—	—	0.035	—
Sulfamethoxazole	0.4	0.005	0.23	0.57	0.14	0.19	0.28	0.23
Number of compounds detected	29	23	23	24	21	25	23	24

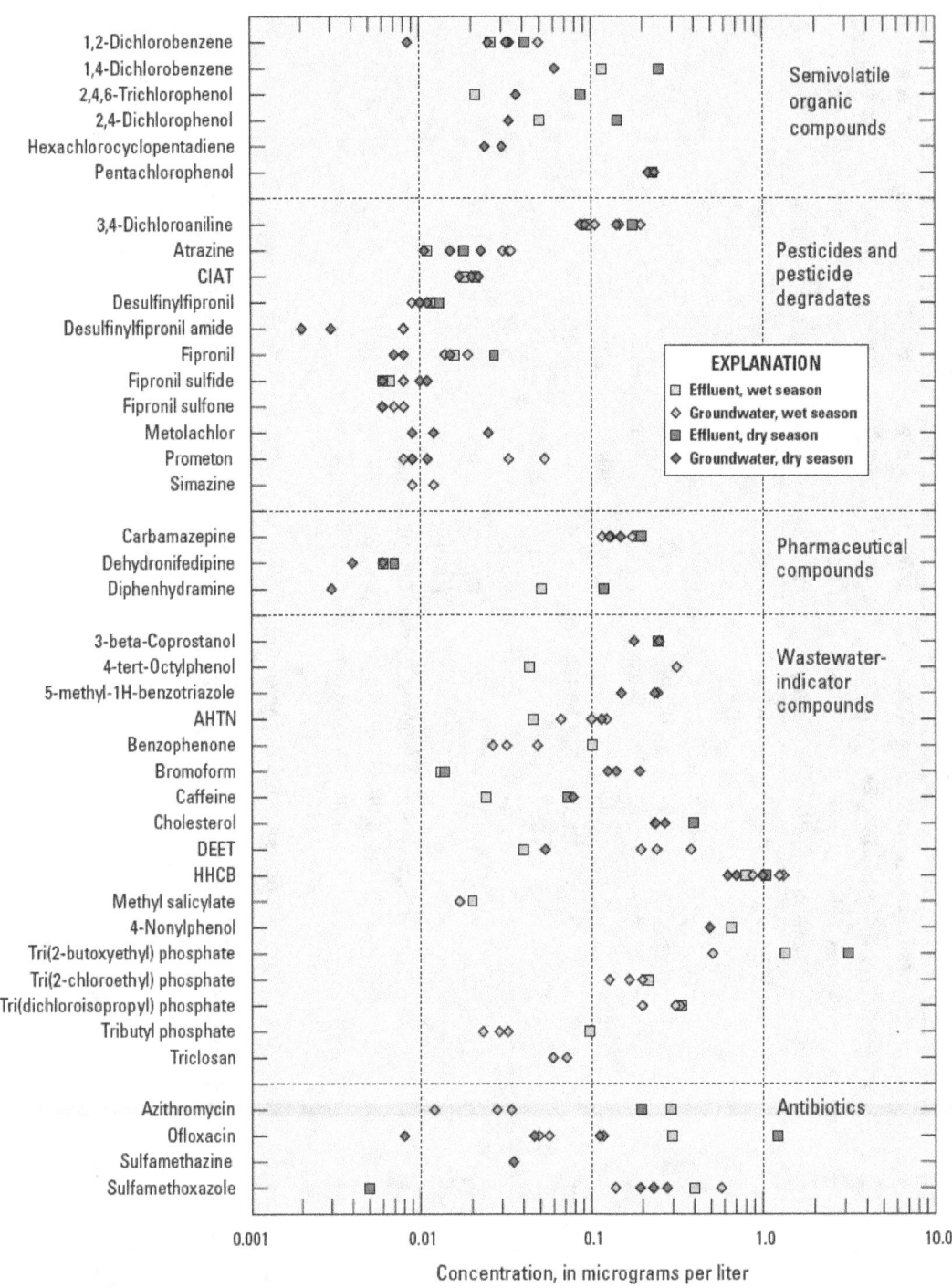

Figure 5. Concentrations of pharmaceuticals and other organic wastewater compounds detected in the Homestead Wastewater Treatment Plant (HSWWTP) effluent and groundwater samples collected from three monitoring wells at the HSWWTP in Florida.

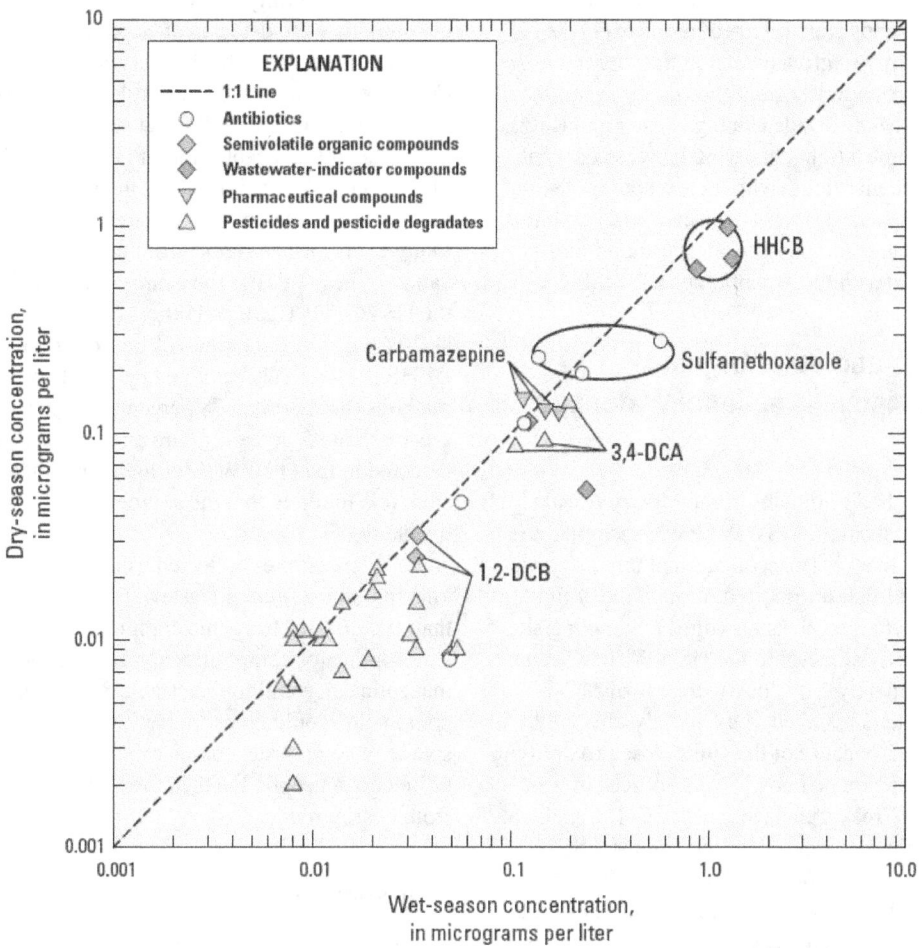

Figure 6. Comparison of pharmaceuticals and other organic wastewater compounds in groundwater samples collected in both the wet season and dry season from three monitoring wells at the Homestead Wastewater Treatment Plant in Florida.

in the Biscayne aquifer (Lindsey and others, 2009) and in public-supply wells near the HSWWTP (Bradner and others, 2005; Foster and Katz, 2010). The presence of fipronil sulfone and desulfinylfipronil amide in groundwater samples could be a result of the oxidation of fipronil and desulfinylfipronil, respectively, after being introduced to the subsurface (Ngim and Crosby, 2001). Both fipronil (E 0.016 and E 0.027 µg/L) and desulfinylfipronil (0.012 and 0.013 µg/L) were detected in effluent composite samples from the HSWWTP. Sulfamethazine is a sulfonamide antimicrobial approved strictly for use in veterinary medicine and has been detected in groundwater near known sources of animal waste (Batt and others, 2006). The detection of sulfamethazine in groundwater from well MW-2 in the dry season could be from an animal waste source (for example, application of animal manure to agricultural fields) near the treatment plant.

More compounds were detected in groundwater samples in the dry season (31 compounds were detected in at least one monitoring well, 15 of which were detected only in the dry season) than in the wet season (26 compounds were detected in at least one monitoring well, 10 of which were detected only in the wet season). When it was possible to compare concentrations of a compound in both the wet season and dry season in a monitoring well, concentrations were generally higher in the wet season (table 7; fig. 6). The higher concentrations in groundwater could be a result of larger effluent loads (pounds per day) during wetter months, which could result in an increased total mass discharge to the Biscayne aquifer with corresponding decreased relative dilution from the native groundwater. However, the cause of this difference is uncertain as the lack of multiple samples in both the wet and dry seasons prevented the opportunity to perform a comparison.

Among the various groups of compounds, the most diagnostic of effluent in the groundwater at this site were HHCB, sulfamethoxazole, 3,4-DCA, 1,2-DCB, and carbamazepine (fig. 6). These compounds were selected because: (1) they were detected frequently in groundwater at this site (table 7) and not detected in groundwater from the Biscayne aquifer in other areas within Miami-Dade County (Foster and Katz, 2010); (2) the data show a tight clustering between effluent and groundwater concentrations (with the exception of sulfamethoxazole) (fig. 5); and (3) they were detected in both the effluent and groundwater samples at concentrations substantially above the MRL (with the exception of 1,2-DCB).

Pharmaceuticals and Other Organic Wastewater Compounds in Canal Water and Bed Sediment

Canal water and bed sediments in the Mowry Canal (MC) near and at a distance from the HSWWTP were sampled to assess the possibility that pharmaceuticals and other organic wastewater compounds are transported through the groundwater system and discharged at canals (fig. 1). Sampling site MC-1 is located directly adjacent to the HSWWTP, whereas site MC-2 is approximately 1 mi "downstream" of MC-1 (fig. 2). The MC-1 site is assumed to be directly affected by the HSWWTP effluent because of the site's close proximity to the plant (fig. 2). There are no known point sources of wastewater to the Snapper Creek Canal (SC) near SC-1. Canal water samples and bed sediments were collected from site SC-1 on one occasion on July 9, 2008, and from sites MC-1 and MC-2 on two occasions on August 12, 2008 (wet season), and December 9, 2009 (dry season). The occurrence and concentrations of selected organic compounds in canal water and sediments from the Mowry and Snapper Creek Canals were compared during this study. The occurrence and concentrations of selected organic compounds in the dry season versus the wet season in canal water and bed sediments from MC-1 and MC-2 are described herein.

Canal Water

Samples were analyzed for the pharmaceuticals and other organic wastewater compounds listed in table 2. A total of 51 compounds (including 28 wastewater-indicator compounds, 7 SVOCs, 12 pesticides and pesticide degradates, 2 pharmaceuticals, 1 antibiotic, and the hormone E2) were detected in one or more canal samples (table 8). Generally, concentrations were below 1.0 µg/L with the exception of DEHP from waters at MC-1 (dry season, E 11 µg/L).

In the Mowry Canal, more compounds were detected in canal water samples collected in the dry season (MC-1, 37; MC-2, 25) than in the wet season (MC-1, 11; MC-2, 8) (table 8), which could indicate that increased groundwater discharge and overland flow during precipitation events increases dilution during the wet season. Furthermore, when

it was possible to compare concentrations of a compound in both the wet and dry seasons at a particular site, concentrations were generally higher in the dry season (fig. 7). Some compounds were detected at concentrations several orders of magnitude higher in the dry season than in the wet season. Only one compound (cholesterol, MC-2) was detected at a higher concentration in the wet season.

The high concentration of DEHP, a widely used plasticizer and inert ingredient in pesticide products, detected in a water sample collected from MC-1 (dry season) is most likely from surface runoff from nearby roads and(or) the application of pesticides to the nearby canal banks, and not from the HSWWTP. Clara and others (2010) detected DEHP as high as 24 µg/L in surface runoff from roads. Furthermore, because DEHP is effectively removed by (most) wastewater-treatment methods (greater than 78 percent, Dargnat and others, 2009; greater than 95 percent, Clara and others; 2010) and it was not detected in the HSWWTP effluent on either sampling occasion, it is unlikely that the source of DEHP to the canal was the HSWWTP.

Water samples collected from SC-1, the site that has no known point source of wastewater, contained low levels (less than E 0.76) of 11 organic compounds, including 5 wastewater-indicator compounds and the non-prescription pharmaceutical acetaminophen (table 8). The detection of these compounds could be an indication that there is a nonpoint source of wastewater to this canal, such as leaky sewer lines, effluent from septic tanks in the area, and(or) surface runoff from nearby roads.

Bed Sediment

Many wastewater-indicator compounds have moderate to large log octanol-water partitioning coefficients and will undergo hydrophobic partitioning into organic-rich bed sediments once released into the environment. Bed sediments were analyzed for the 57 wastewater-indicator compounds listed in table 3 and are the same wastewater-indicator compounds analyzed in water samples. Previous studies have used the presence of these compounds in bed sediments as a diagnostic tool for assessing the discharge of wastewater into surface-water bodies (Burkhardt and others, 2006; Moon and others, 2008).

Bed sediments from all three sites included in this study contained a variety of wastewater-indicator compounds (fig. 8; table 9). For ease of comparison and discussion, the wastewater-indicator compounds included in this section have been grouped into three general categories: (1) personal care products (PCPs), (2) biogenic sterols, and (3) others (for example, PAHs and alkyl-PAHs, wood preservative, industrial solvent,and so forth). The compounds that comprise each of these groups are listed in table 3.

The sediments from the two Mowry Canal sites had the same number (17) of detections of wastewater-indicator compounds, 15 of which were detected at both sites (table 9). More compounds were detected in sediments at MC-1 (15), the site directly next to the HSWWTP, than at MC-2 (12),

Table 8. Pharmaceuticals and other organic wastewater compounds detected in canal-water samples.

[Values in micrograms per liter. SC-1, Snapper Creek Canal Site 1; SC-1-R, Snapper Creek Canal Site 1 Replicate; MC-1, Mowry Canal Site 1; MC-2, Mowry Canal Site 2; IMB, imbalance; NWQL, U.S. Geological Survey National Water Quality Laboratory; CLLE, Continuous liquid-liquid extraction; GC/MS, gas chromatograpahy/mass spectrometry; OGRL, U.S. Geological Survey Organic Geochemistry Research Laboratory; SPE, solid-phase extraction; HPLC/MS high-performance liquid chromatography/mass spectrometry; ELISA, enzyme-linked immunosorbent assay; —, not detected; E, estimated value]

Compound	SC-1	SC-1-R	MC-1	MC-1	MC-2	MC-2
Site name	SC-1	SC-1-R	MC-1	MC-1	MC-2	MC-2
Sample date (yyyymmdd)	20080709	20080709	20080812	20091209	20080812	20091209
Sample time	900	905	1200	1130	1600	1230
Season	Wet	Wet	Wet	Dry	Wet	Dry
Semivolatile organic compounds (NWQL, Unfiltered, CLLE, GC/MS)						
1,2,4-Trichlorobenzene	—	—	E 0.017	—	—	—
1,2-Dichlorobenzene	—	—	E 0.012	—	—	—
2,4-Dichlorophenol	E 0.014	—	—	—	—	—
Diethyl phthalate	—	—	—	E 0.13	—	—
Diethylhexyl phthalate (DEHP)	—	—	—	E11	—	—
Di-n-octyl phthalate	E 0.76	—	—	—	—	—
Pentachlorophenol	—	—	E 0.20	—	—	—
Pesticides and pesticide degradates (NWQL, Filtered, SPE, GC/MS)						
2-Chloro-4-isopropylamino-6-amino-s-triazine (CIAT)	E 0.006	E 0.006	E 0.011	E 0.02	E 0.009	E 0.022
3,4-dichloroaniline	—	—	—	E 0.021	—	E 0.01
Atrazine	0.028	0.028	0.011	0.066	0.01	0.075
Desulfinylfipronil	—	—	—	E 0.014	—	E 0.013
Dichlorvos	—	—	0.35	—	0.46	—
Fipronil	—	—	—	E 0.009	—	—
Fipronil sulfide	—	—	—	E 0.007	—	—
Hexazinone	0.035	0.038	—	—	—	—
Metolachlor	—	—	E 0.008	0.6	E 0.007	0.7
Prometon	—	—	E 0.005	E 0.008	—	E 0.008
Simazine	—	—	—	—	E 0.004	—
Terbuthylazine	—	—	—	0.012	—	0.012
Wastewater-indicator compounds (NWQL, Unfiltered, CLLE, GC/MS)						
2,6-Dimethylnaphthalene	E 0.006	E 0.006	—	E 0.003	—	—
3-*beta*-Coprostanol	—	—	—	E 0.27	—	—
4-Nonylphenol diethoxylate (NP$_2$EO)	—	—	—	E 0.45	—	E 0.42
4-Nonylphenol monoethoxylate (NP$_1$EO)	—	E 0.22	—	—	—	—
4-Octylphenol diethoxylate (OP$_2$EO)	—	—	—	E 0.21	—	E 0.41
4-Octylphenol monoethoxylate (OP$_1$EO)	—	—	—	E 0.18	—	E 0.17
4-tert-Octylphenol (OP)	—	—	—	—	—	E 0.034

Table 8. Pharmaceuticals and other organic wastewater compounds detected in canal-water samples.—Continued

[Values in micrograms per liter. SC-1, Snapper Creek Canal Site 1; SC-1-R, Snapper Creek Canal Site 1 Replicate; MC-1, Mowry Canal Site 1; MC-2, Mowry Canal Site 2; IMB, imbalance; NWQL, U.S. Geological Survey National Water Quality Laboratory; CLLE, Continiuous liquid-liquid extraction; GC/MS, gas chromatogrpahy/mass spectrometry; OGRL, U.S. Geological Survey Organic Geochemistry Research Laboratory; SPE, solid-phase extraction; HPLC/MS high-performance liquid chromatogrpahy/mass spectrometry; ELISA, enzyme-linked immunosorbent assay; —, not detected; E, estimated value]

Compound	SC-1 20080709 900 Wet	SC-1-R 20080709 905 Wet	MC-1 20080812 1200 Wet	MC-1 20091209 1130 Dry	MC-2 20080812 1600 Wet	MC-2 20091209 1230 Dry
5-Methyl-1H-benzotriazole	—	—	—	E 0.072	—	—
Benz[a]pyrene	—	—	—	E 0.007	—	—
beta-Sitosterol	—	—	—	E 0.33	—	E 0.23
beta-Stigmastanol	E 0.62	E 0.5	—	—	—	—
Bromacil	—	—	—	E 0.20	—	E 0.10
Caffeine	E 0.027	E 0.023	—	E 0.009	—	—
Cholesterol	E 0.72	E 0.59	—	E 0.52	E 0.72	E 0.27
Fluoranthene	—	—	—	E 0.013	—	—
Golaxolide (HHCB)	—	—	—	E 0.089	—	E 0.036
Isophorone	—	—	E 0.015	—	E 0.015	—
Menthol	—	—	—	E 0.018	—	—
N,N-diethyl-meta-toluamide (DEET)	—	—	—	E 0.027	—	E 0.013
Naphthalene	—	—	—	E 0.01	—	E 0.011
Pyrene	—	—	—	E 0.008	—	—
Tonalide (AHTN)	—	—	—	E 0.008	—	—
Tri(2-butoxyethyl) phosphate (TBEP)	—	—	—	E 0.30	E 0.076	E 0.11
Tri(2-chloroethyl) phosphate (TCEP)	—	—	—	E 0.084	—	E 0.042
Tri(dichloroisopropyl) phosphate (TPIP)	—	—	E 0.029	E 0.16	—	E 0.075
Tributyl phosphate (TBP)	—	—	—	E 0.027	—	E 0.012
Triclosan	—	—	—	E 0.038	—	E 0.043
Triphenyl phosphate	—	—	—	E 0.01	—	E 0.007
Pharmaceutical compounds (NWQL, Filtered, SPE, HPLC/MS)						
Acetaminophen	E 0.008	E 0.008	—	—	—	—
Cabamazepine	—	—	—	0.057	—	0.055
Antibiotics (OGRL, Filtered, SPE, LC/MS)						
Sulfamethoxazole	—	—	—	0.11	—	0.12
Hormone (IMB)						
17-beta-Estradiol (E2)	—	—	0.002	0.005	—	—
Number of compounds detected	10	9	11	37	8	25

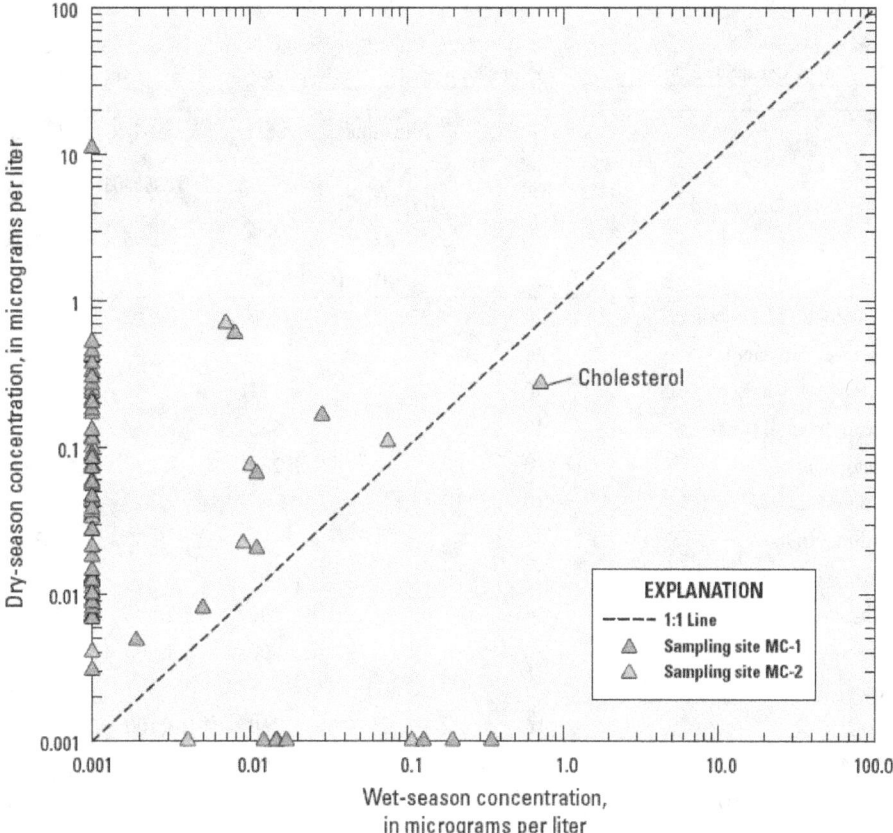

Figure 7. Comparison of wet- and dry-season concentrations of pharmaceuticals and other organic wastewater compounds detected in canal-water samples collected from canal sampling sites MC-1 and MC-2 in Florida. Concentrations reported by the laboratory as less than the method reporting level are plotted as 0.001 micrograms per liter.

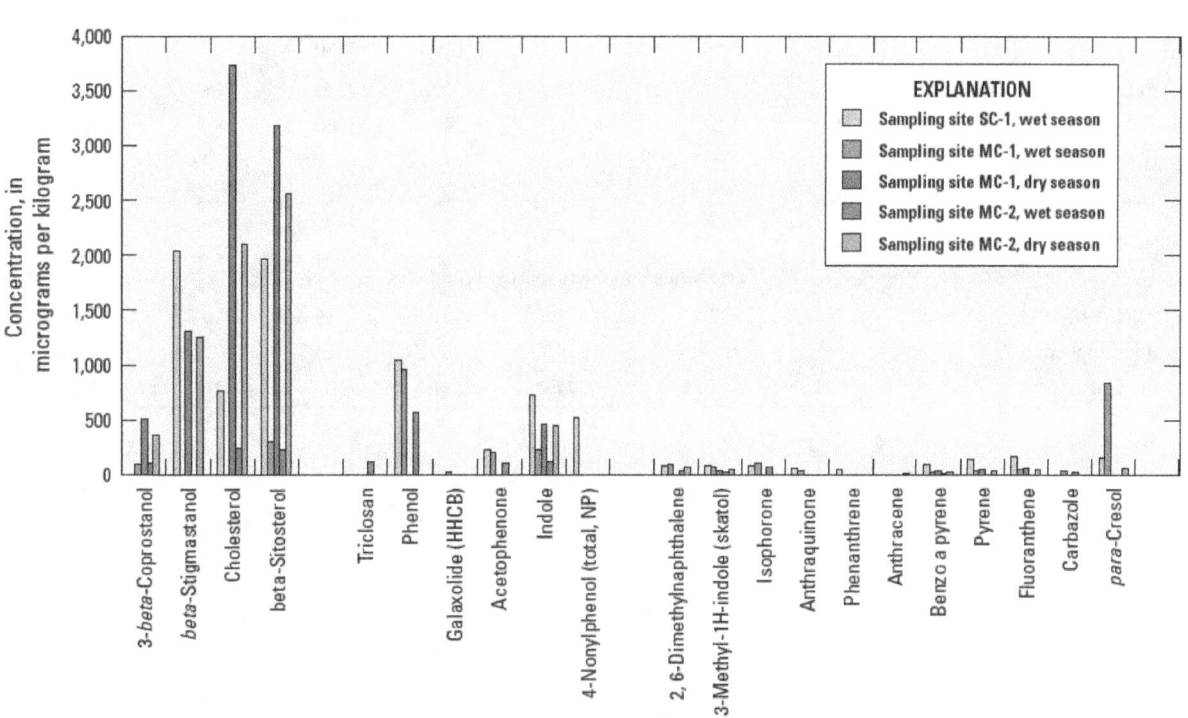

Figure 8. Concentrations of wastewater-indicator compounds in bed sediment samples collected from canal sampling sites SC-1, MC-1, and MC-2.

Table 9. Wastewater-indicator compounds detected in canal bed sediment samples.

[Values in micrograms per kilogram (µg/kg). SC-1, Snapper Creek Canal Site 1; SC-1-R, Snapper Creek Canal Site 1 Replicate; MC-1, Mowry Canal Site 1; MC-2, Mowry Canal Site 2; MRL, method reporting level; —, not detected; E, estimated]

Compound	MRL (µg/kg)	Site name	SC-1	SC-1-R	MC-1	MC-1	MC-2	MC-2
		Sample date (yyyymmdd)	20080709	20080709	20080812	20091209	20080812	20091209
		Sample time	0900	0905	1200	1130	1600	1230
		Season	Wet	Wet	Wet	Dry	Wet	Dry
2,6-Dimethylnaphthalene	24		86	110	97	—	E 33	E 70
3-*beta*-Coprostanol	36		—	—	E 89	E 500	E 100	E 360
3-Methyl-1H-indole (Skatol)	30		81	86	70	E 38	E 20	E 47
4-Nonylphenol (total)	49		E 520	—	—	—	—	—
Acetophenone	10		E 230	E 220	E 200	—	E 110	—
Anthracene	19		—	—	—	—	—	E 6.0
Anthraquinone	24		—	E 57	E 32	—	—	—
Benzo[a]pyrene	24		94	120	E 17	E 32	E 12	E 21
beta-Sitosterol	36		E 2,000	E 2,500	E 300	E 3,200	E 230	E 2,600
beta-Stigmastanol	36		E 2,000	E 1,900	—	E 1,300	—	E 1,300
Carbazole	22		—	—	E 27	—	E 15	—
Cholesterol	16		E 760	E 1,100	—	E 3,700	E 230	E 2,100
Fluoranthene	23		160	190	46	E 4	—	E 48
Galaxolide (HHCB)	16		—	—	E 18	—	—	—
Indole	53		720	1,100	E 230	E 460	E 120	E 440
Isophorone	43		E 83	E 67	E 100	—	E 64	—
para-Cresol	16		E 150	E 140	830	—	—	E 56
Phenanthrene	20		E 45	E 50	—	—	—	—
Phenol	38		E 1,000	E 740	E 960	—	E 570	—
Pyrene	20		150	170	E 37	E 43	—	E 38
Triclosan	49		—	—	—	—	110	—
Number of OWCs detected			15	15	15	9	12	12
Surrogate compounds (percent recovery)								
Bisphenol A-d_3			51.2	51.3	14.2	58.7	21.4	60.3
Decafluorobiphenyl			30.7	30.2	13.7	26.6	22.4	26.0
Fluoranthene-d_{10}			74.0	71.4	102	169	79.7	187

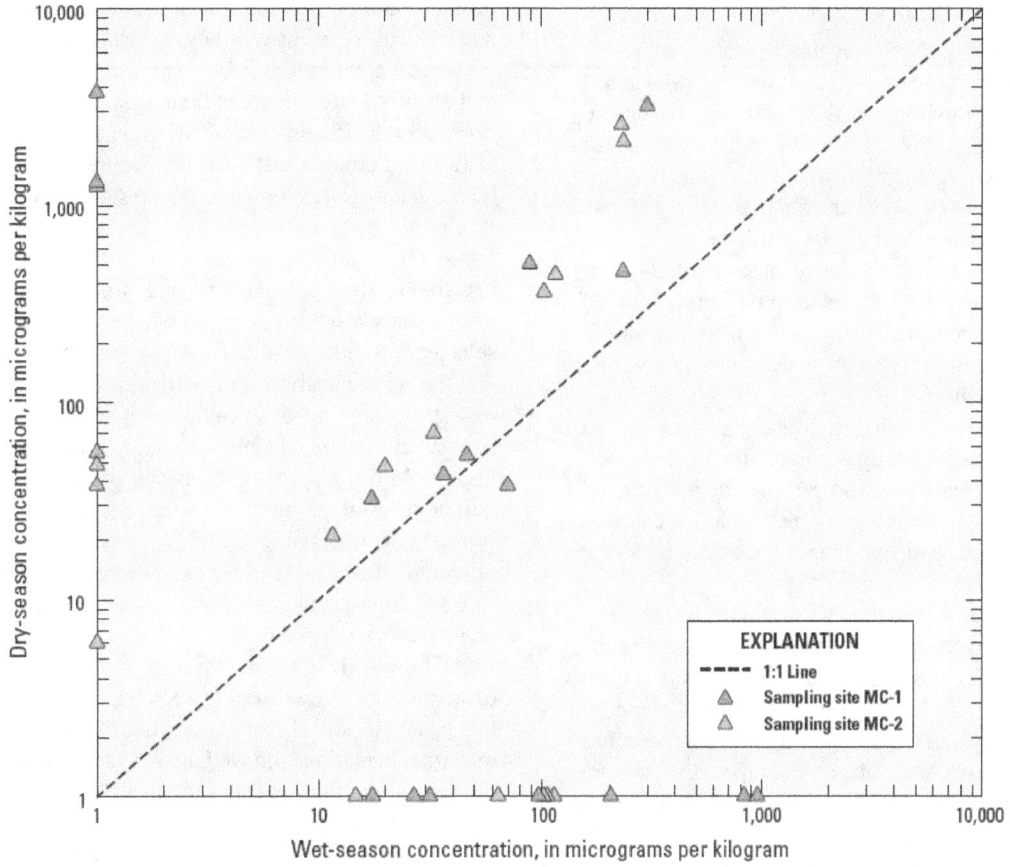

Figure 9. Comparison of wet- and dry-season concentrations of organic wastewater compounds in bed sediment samples collected from canal sampling sites MC-1 and MC-2 in Florida. Concentrations reported by the laboratory as less than the method reporting level are plotted as 1 microgram per kilogram.

the site approximately 1 mi downstream, in the wet season. Conversely, more compounds were detected in sediments at MC-2 (12) than at MC-1 (9), in the dry season. The four biogenic sterols were detected at high concentrations (ranging from 89 to 3,700 mg/kg) in bed sediments from both locations. Phenol was also detected at high concentrations (960 and 570 mg/kg) in bed sediments collected in the wet season.

Similar to canal water results, when it was possible to compare concentrations of a compound in bed sediments in both the wet and dry season at a particular site, concentrations were generally higher in the dry season (fig. 9; table 9). Some compounds were detected several orders of magnitude higher in the dry season than in the wet season. The higher concentrations in bed sediment in the dry season could be due to greater deposition of particulates during low-flow conditions in the dry season; however, the three surrogate compounds (bisphenol A-d_3, decafluorobiphenyl, and fluoranthene-d_{10}) had lower percent recoveries in the wet-season samples compared to the dry-season samples (table 9), thus indicating the extraction

method did not perform as well on the wet-season samples. This result could partially explain the lower concentrations of wastewater-indicator compounds in bed sediments collected in the wet season. Only one compound (3-methyl-1H-indole (skatol), MC-1) was detected at a higher concentration in the wet season.

Similar to the water samples collected from SC-1, bed sediment collected at SC-1 contained detectable quantities of 16 wastewater-indicator compounds (fig. 8; table 9). Some of the compounds detected in the bed sediment sample at SC-1 (including 3-methyl-1H-indole and the four biogenic sterols— 3-*beta*-coprostanol, *beta*-sitosterol, *beta*-stigmastanol and cholesterol) do have non-anthropogenic sources, which can partially explain the presence of these compounds. However, the presence of other compounds, such as 4-nonylphenol and isophorone, which are strictly anthropogenic, as well as the 11 compounds detected in water samples (mentioned in the previous discussion), further suggests the presence of a nonpoint source of wastewater to the canal.

Summary and Conclusions

The U.S. Geological Survey in cooperation with the Miami-Dade Water and Sewer Department (WASD) and the Department of Environmental Resources Management (DERM), initiated a study to assess the presence of pharmaceuticals and other organic wastewater compounds in influent and effluent at four wastewater-treatment plants (WWTPs) in Miami-Dade County during 2008 and 2009. A Lagrangian sampling scheme was used at each plant to collect 24-hour (hr) flow-weighted composite samples from the influent and effluent. Samples were collected at each plant in both the dry (low-flow) and wet (high-flow) seasons to determine any seasonal variations in concentrations.

Water samples were analyzed for approximately 210 organic compounds including: semivolatile organic compounds (SVOCs), pesticides and pesticide degradates, wastewater-indicator compounds, prescription and non-prescription pharmaceuticals, antibiotics, and one hormone (17-*beta*-estradiol).

Influent samples at each plant contained a complex mixture of organic compounds reflecting the diversity of incoming domestic, municipal, and industrial waste. Compounds detected in 24-hr flow-weighted influent composite samples included: 20 SVOCs, 12 pesticides and pesticide degradates, 52 wastewater-indicator compounds, 5 pharmaceuticals, 14 antibiotics, and the hormone 17-*beta*-estradiol. Wastewater-indicator compounds accounted for nearly all of the total concentrations in each influent sample collected. Two biogenic sterols, cholesterol and 3-*beta*-coprostanol, had the highest median concentrations (91 and 79 micrograms per liter (µg/L), respectively), while diethylhexyl phthalate, a widely used plasticizer and inert ingredient in pesticide products, showed the greatest concentration range (5.3 to 270 µg/L). The total concentrations in influent samples collected in the dry season were generally 10 to 30 percent higher than samples collected in the wet season.

Compounds detected in 24-hr flow-weighted effluent composite water samples included: 19 SVOCs, 13 pesticides, 49 wastewater-indicator compounds, 7 pharmaceuticals, 11 antibiotics, and 17-*beta*-estradiol. Concentrations of individual organic compounds in effluent samples were comparable with other reported values in WWTP effluent across the United States. Using total concentration in each effluent sample and average flows during sampling; estimated effluent loads were calculated and ranged from 0.3 to 25.7 pounds per day. Wastewater-indicator compounds accounted for greater than 64 percent of the total load at each plant. Loads were generally higher in the wet season at each plant, reflecting the higher flows during sample collection. Among the various groups of compounds, SVOCs and wastewater-indicator compounds generally had higher removal efficiencies than pharmaceuticals, antibiotics and pesticides.

The number of samples collected at each plant was small; only two samples were collected at each plant. The small number of samples was adequate for the primary objectives of this study, which were to document the occurrence of pharmaceuticals and other organic wastewater compounds in influent and effluent waters at WWTPs in the county and to determine any seasonal variations in concentrations, percent reduction, and effluent loads. The results suggest there are seasonal variations in both effluent concentrations and loads. To determine if these variations are statistically different, future studies will be needed to collect multiple samples during both seasons. It might be more efficient to analyze samples for compounds that are frequently detected in effluent waters and exclude less frequently detected compounds from target lists. Results of this study provide a basis on which compounds could be selected in follow-up studies. Analytical chemists are continually improving analytical methods for the low-level detection of organic compounds in complex environmental matrices. As extraction methods improve and analytical detection levels become lower, an even broader range of organic compounds will be detectable. Future occurrence studies should consider if these new methods are applicable. Additionally, influent and effluent matrices are complex and highly variable; therefore any future studies should consider including a rigorous quality-assurance plan.

The WASD is currently seeking to increase the amount of wastewater reused from its regional wastewater-treatment plants, and is actively pursuing reuse plans for aquifer recharge, irrigation, and wetland rehydration. To investigate the fate of pharmaceuticals and other organic wastewater compounds in groundwater affected by WWTP effluent, water samples were collected from three monitoring wells at the Homestead WWTP in both the wet and dry seasons. For decades, this plant has been discharging tertiary-treated effluent directly to the water table. Generally, concentrations in the groundwater were below 0.5 µg/L with the exception of the fragrance, galaxolide (as high as 1.3 µg/L), the antibiotic, sulfamethoxazole (as high as 0.57 µg/L), and the flame retardant, tri(2-butoxyethyl)phosphate (detected in one sample at E 0.51 µg/L). The attenuation of pharmaceuticals and other organic wastewater compounds in groundwater at the HSW-WTP was evident by lower concentrations in groundwater samples compared with effluent samples, likely from biodegradation and dilution with recharge water from upgradient locations. Nine compounds were detected in groundwater samples that were not detected in effluent samples. The presence of these compounds in groundwater at this site could be from chemical reactions occurring in the subsurface after disposal or possibly from sources other than the WWTP. Concentrations of pharmaceuticals and other organic wastewater compounds were generally higher in groundwater samples collected in the wet season, which most likely reflects the higher loads of these compounds in the effluent during the wet season. A more in-depth geochemical study, one in which the three monitoring wells at the plant and other monitoring wells further downgradient of the HSWWTP are sampled and analyzed for major ions, trace metals, and dissolved gasses as well as the isotopic analysis of oxygen-, carbon-, hydrogen-, and nitrogen-bearing materials, may be worthwhile to delineate the plume and help to understand the ultimate fate of these compounds as they are transported offsite.

Surface-water and bed sediments samples were collected from two canal sites near the Homestead WWTP to determine if constituents originating from the effluent could be transported from the groundwater to the canal. These locations were sampled once in the dry season and once in the wet season to determine any seasonal variations. A total of 51 compounds were detected in one or more canal water sample collected at concentrations generally below 1.0 µg/L. Results from seasonal sampling from the two canals sites show concentrations of pharmaceuticals and other organic wastewater compounds were generally higher in the dry season.

Detection of pharmaceuticals and other organic wastewater compounds in water and sediment samples collected at SC-1 suggests that even away from wastewater-treatment facilities, these compounds may be present in canals from unknown sources, thus it may be difficult to link organic wastewater compounds in surface water to a municipal wastewater sources. A more detailed water- quality study, one in which multiple canal sites throughout the study area are sampled, might help resolve the source of these compounds to the canals, or at least identify areas where these compounds are more frequently detected.

References

Andersen, H., Siegrist, H., Halling-Sorensen, B., and Ternes, T.A., 2003, Fate of estrogens in a municipal sewage treatment plant: Environmental Science and Technology, v. 37, no.18, p. 4021–4026.

Barnes, K.K., Christenson, S.C., Kolpin, D.W., Focazio, M.J., Furlong, E.T., Zaugg, S.D., Meyer, M.T., and Barber, L.B., 2004, Pharmaceuticals and other waste water contaminants within a leachate plume downgradient of a municipal landfill: Ground Water Monitoring and Remediation, v. 24, no. 2, p. 119–126.

Barber, L.B., Lee, K.E., Swackhamer, D.L., Schoenfuss, H.L., 2007, Reproductive responses of male fathead minnows exposed to wastewater treatment plant effluent, effluent treated with XAD8 resin, and an environmentally relevant mixture of alkylphenol compounds: Aquatic Toxicology, v. 82, no. 1, p. 36-46, doi:10.1016/j.aquatox.2007.01.003.

Barnes, K.K., Kolpin, D.W., Focazio, M.J., Furlong, E.T., Meyer, M.T., Zaugg, S.D., Haack, S.K., Barber, L.B., and Thurman, E.M., 2008, Water-quality data for pharmaceuticals and other organic wastewater contaminants in ground water and in untreated drinking water sources in the United States, 2000–01: U.S. Geological Survey Open-File Report 2008–1293, 7 p. plus tables.

Batt, A.L., Snow, D.D., and Aga, D.S., 2006, Occurrence of sulfonamide antimicrobials in private water wills in Washington County, Idaho, USA: Chemosphere, v. 64, no. 11, p. 1963–1971.

Bradner, A., McPherson, B.F., Miller, R.L., Kish, G., and Bernard, B., 2005, Quality of ground water in the Biscayne aquifer in Miami-Dade, Broward, and Palm Beach Counties, Florida, 1996-1998, with Emphasis on Contaminants: U.S. Geological Survey Open-File Report 2004-1438, 20 p.

Burkhardt, M.R., Zaugg, S.D., Smith, S.G., and ReVello, R.C., 2006, Determination of wastewater compounds in sediment and soil by pressurized solvent extraction, solid-phase extraction, and capillary-column gas chromatography/mass spectrometry: U.S. Geological Survey Techniques and Methods, book 5, chap. B2, 40 p.

Cahill, J.D., Furlong, E.T., Burkhardt, M.R., Kolpin, D.W., and Anderson, L.G., 2004, Determination of pharmaceutical compounds in surface- and ground-water samples by solid-phase extraction and high-performance liquid chromatography--electrospray ionization mass spectrometry: Journal of Chromatography A, v. 1041, no. 1-2, p. 171–180, doi: 10.1016/j.chroma.2004.04.005.

Clara, M., Strenn, B., and Kreuzinger, N., 2004, Carbamazepine as a possible anthropogenic marker in the aquatic environment: investigations on the behavior of carbamazepine in wastewater treatment and during groundwater infiltration: Water Research, v. 38, p. 947–954.

Clara, M., Windhofer, G., Hartl, W., Braun, K., Simon, M., Gans, O., Scheffknecht, C., and Chovanec, A., 2010, Occurrence of phthalates in surface runoff, untreated and treated wastewater and fate during wastewater treatment: Chemosphere, v. 78, no. 9, p. 1078–1084.

Cordy, G.E., Duran, N.L., Bouwer, H., Rice, R.C., Furlong, E.T., Zaugg, S.D., Meyer, M.T., Barber, L.B., and Kolpin, D.W., 2004, Do pharmaceuticals, pathogens, and other organic waste water compounds persist when waste water is used for recharge?: Ground Water Monitoring and Remediation, v. 24, no. 2, p. 58–69.

Cunningham, K.J., Carlson, J.L., Wingard, G.L., Robinson, E., and Wacker, M.A., 2004, Characterization of aquifer heterogeneity using cyclostratigraphy and geophysical methods in the upper part of the karstic Biscayne aquifer, Southeastern Florida: U.S. Geological Survey Water-Resources Investigations Report 03-4208, 66 p., 5 apps. (on CD), and 5 pls.

Dargnat C., Teil M., Chevreuil M., and Blanchard M., 2009, Phthalate removal throughout wastewater treatment plant Case study of Marne Aval station (France): Science of the Total Environment, v. 407, p. 1235–1244.

Farre, M., Brix, R., Kuster, M., Rubio, F., Goda, Y., Lopez de Alda, M.J., and Barcelo, D., 2006, Evaluation of commercial immunoassays for the detection of estrogens in water by comparison with high-performance liquid chromatography tandem mass spectrometry HPLC–MS/MS (QqQ): Analytical and Bioanalytical Chemistry, v. 385, p. 1001-1011.

Fish, J.E., and Stewart, M.T., 1991, Hydrogeology of the surficial aquifer system, Dade County, Florida: U.S. Geological Survey Water-Resources Investigations Report 90-4108, 56 p.

Fishman, M.J., ed., 1993, Methods of analysis by the U.S. Geological Survey National Water Quality Laboratory--Determination of inorganic and organic constituents in water and fluvial sediments: U.S. Geological Survey Open-File Report 93-125, 217 p.

Florida Department of Environmental Protection, 2010, 2008 Reuse Inventory: Tallahassee, Florida Department of Environmental Protection.

Foster, A.L., and Katz, B.G., 2010, Occurrence of organic compounds in source and finished samples from seven drinking-water treatment facilities in Miami-Dade County, Florida, 2008: U.S. Geological Survey Data Series 550, 22 p.

Furlong, E.T., Werner, S.L., Anderson, B.D., and Cahill, J.D., 2008, Methods of analysis by the U.S. Geological Survey National Water Quality Laboratory determination of human-health pharmaceuticals in filtered water by chemically modified styrene-divinylbenzene resin-based solid-phase extraction and high-performance liquid chromatography/mass spectrometry: U.S. Geological Survey Techniques and Methods, book 5, sec. B, chap. B5, 56 p.

Heberer, T., Mechlinski, A., Fanck, B., Knappe, A., Massmann, G., Pekdeger, A., and Fritz, B., 2004, Field studies on the fate and transport of pharmaceutical residues in bank filtration: Ground Water Monitoring and Remediation, v. 24, no. 2, p. 70–77.

Kinney, C.A., Furlong, E.T., Werner, S.L., and Cahill, J.D., 2006, Presence and distribution of wastewater-derived pharmaceuticals in soil irrigated with reclaimed water: Environmental Toxicology and Chemistry, v. 25, no. 2, p. 317–326, doi: 10.1897/05-187R.1.

Kolpin, D.W., Furlong, E.T., Meyer, M.T., Thurman, E.M., Zaugg, S.D., Barber, L.B., and Buxton, H.T., 2002, Pharmaceuticals, hormones, and other organic wastewater constituents in U.S. streams, 1999–2000: A national reconnaissance: Environmental Science and Technology, v. 36, no. 6, p. 1202–1211.

Lietz, A.C., and Meyer, M.T., 2006, Evaluation of emerging contaminants of concern at the South District Wastewater Treatment Plant based on seasonal events, Miami-Dade County, Florida, 2004: U.S. Geological Survey Scientific Investigations Report 2006-5240, 38 p.

Lindsey, B.D., Berndt, M.P., Katz, B.G., Ardis, A.F., and Skach, K.A., 2009, Factors affecting water quality in selected carbonate aquifers in the United States, 1993–2005: U.S. Geological Survey Scientific Investigations Report 2008-5240, 117 p.

Meyer, J., and Bester, K., 2004, Organophosphate flame retardants and plasticisers in wastewater treatment plants: Journal of Environmental Monitoring, v. 6, p. 599–605.

Meyer, M.T., Lee, E.A., Ferrell, G.M., Bumgarner, J.E., and Varns, J., 2007, Evaluation of offline tandem and online solid-phase extraction with liquid chromatography/electrospray ionization-mass spectrometry for analysis of antibiotics in ambient water and comparison to an independent method: U.S. Geological Survey Scientific Investigations Report 2007-5021, 28 p.

Moon, H., Yoon, S., Jung, R., and Choi, M., 2008, Wastewater treatment plants (WWTPs) as a source of sediment contamination by toxic organic pollutants and fecal sterols in a semi-enclosed bay in Korea: Chemosphere, v. 73, p. 880–889.

National Oceanic and Atmospheric Agency (NOAA), 2004, Climatography of the United States No. 20 Monthly Station Climate Summaries for the 1971-2000: National Climatic Data Center, accessed 2011 at *http://www.ncdc.noaa.gov/ normals.html.*

National Research Council, 1998, Issues in Potable Reuses - The Viability of Augmenting Drinking Water Supplies with Reclaimed Water, National Academy Press: Washington, DC, 1998.

Ngim, K.K., and Crosby, D.G., 2001, Abiotic processes influencing fipronil and desthiofipronil dissipation in California, USA, rice fields: Environmental Toxicology and Chemistry, v. 20, no. 5, p. 972–977.

Pehlivanoglu-Mantas, E., Hawley, E.L., Deeb, R.A., and Sedlak, D.L., 2006, Formation of nitrosodimethylamine (NDMA) during chlorine disinfection of wastewater effluents prior to use in irrigation systems: Water Research, v. 40, p. 341–347.

Phillips, P.J., Smith, S.G., Kolpin, D.W., Zaugg, S.D., Buxton, H.T., Furlong, E.T., Esposito, Kathleen, and Stinson, Beverley, 2010, Pharmaceutical formulation facilities as sources of opioids and other pharmaceuticals to wastewater treatment plant effluents: Environmental Science and Technology, v. 44, no. 13, p. 4910–4916.

Renken, R.A., Cunningham, K.J., Shapiro, A.M., Harvey, R.W., Zygnerski, M.R., Metge, D.W., and Wacker, M.A., 2008, Pathogen and chemical transport in the karst limestone of the Biscayne aquifer: 1. Revised conceptualization of groundwater flow: Water Resources Research, v. 44, 16 p.

Sando, S.K., Furlong, E.T., Gray, J.L., Meyer, M.T., and Bartholomay, R.C., 2005, Occurrence of organic wastewater compounds in wastewater effluent and the Big Sioux River in the Upper Big Sioux River Basin, South Dakota, 2003–2004: U.S. Geological Survey Scientific Investigations Report 2005-5249, 108 p.

Sandstrom, M.W., Stroppel, M.E., Foreman, W.T., and Schroeder, M.P., 2001, Methods of analysis by the U.S. Geological Survey National Water Quality Laboratory--Determination of moderate-use pesticides and selected degradates in water by C-18 solid-phase extraction and gas chromatography/mass spectrometry: U.S. Geological Survey Water-Resources Investigations Report 01-4098, 70 p.

Schultz, M.M., Furlong, E.T., Kolpin, D.W., Werner, S.L., Schoenfuss, H.L., Barber, L.B., Blazer, V.S., Norris, D.O., and Vajda, A.M., 2010, Antidepressant pharmaceuticals in two U.S. effluent-impacted streams--Occurrence and fate in water and sediment, and selective uptake in fish neural tissue: Environmental Science and Technology, v. 44, no.6, p. 1918–1925, doi:10.1021/es9022706

Stackelberg, P.E., Furlong, E.T., Meyer, M.T., Zaugg, S.D., Henderson, A.K., and Reissman, D.B., 2004, Persistence of pharmaceutical compounds and other organic wastewater contaminants in a conventional drinking-water-treatment plant: Science of the Total Environment, v. 329, p. 99–113.

Taylor, J.K., 1987, Quality-assurance of chemical measurements: Chelsea, Mich., Lewis Publishers, 328 p.

U.S. Environmental Protection Agency, 1994, Guidelines for Water Reuse. EPA/625/R-04/108. September 1994, 478 p.

U.S. Geological Survey, 2010, U.S.G.S. Toxic Substances Hydrology Program, 2010: U.S. Geological Survey Fact Sheet 2010-3011. (Also available at *http://pubs.usgs.gov/ fs/2010/3011/.*)

U.S. Geological Survey, variously dated, National field manual for the collection of water-quality data: U.S. Geological Survey Techniques of Water-Resources Investigations, book 9, chaps. A1–A9. (Also available at *http://pubs.water. usgs.gov/twri9A.*)

Vajda, A.M., Barber, L.B., Gray, J.L., Lopez, E.M., Woodling, J.D., and Norris, D.O., 2008, Reproductive disruption in fish downstream from an estrogenic wastewater effluent: Environmental Science and Technology, v. 42, no. 9, p. 3407–3414, doi:10.1021/es0720661.

WaterReuse Foundation, 2008, National database of water reuse facilities summary report. 02-004-01. WaterReuse Foundation, Alexandria, Virginia, 2008

Wilkison, D.H., Armstrong, D.J., Norman, R.D., Poulton, B.C., Furlong, E.T., and Zaugg, S.D., 2006, Water quality in the Blue River basin, Kansas City metropolitan area, Missouri and Kansas, July 1998 to October 2004: U.S. Geological Survey Scientific Investigations Report 2006–5147, 170 p.

Writer, J.H., Barber, L.B., Ryan, J.N., and Bradley, P.M., 2011, Biodegradation and attenuation of steroidal hormones and alkylphenols by stream biofilms and sediments: Environmental Science and Technology, v. 45, no. 10, p. 4370–4376, doi:10.1021/es2000134.

Zaugg, S.D., Sandstrom, M.W., Smith, S.G., and Fehlberg, K.M., 1995, Methods of analysis by the U.S. Geological Survey National Water Quality Laboratory--Determination of pesticides in water by C-18 solid-phase extraction and capillary-column gas chromatography/mass spectrometry with selected-ion monitoring: U.S. Geological Survey Open-File Report 95-181, 60 p.

Zaugg, S.D., Smith, S.G., and Schroeder, M.P., 2006, Determination of wastewater compounds in whole water by continuous liquid-liquid extraction and capillary-column gas chromatography/mass spectrometry: U.S. Geological Survey Techniques and Methods, book 5, chap. B4, 30 p.

Appendixes 1–4

Appendix 1. Quality-assurance data for blank, replicate, and matrix spike samples.

[Recovery values outside expected ranges are in red. MRL, method reporting level; µg/L, micrograms per liter; RPD, Relative Percent Difference; %, percent; NWQL, U.S. Geological Survey National Water Quality Laboratory; CLLE, Continuous liquid-liquid extraction; GC/MS, gas chromatograhy/mass spectrometry; OGRL, U. S. Geological Survey Organic Geochemistry Research Laboratory; SPE, solid-phase extraction; HPLC/MS, high-performance liquid chromatograhy/mass spectrometry; ELISA, enzyme-linked immunosorbent assay; —, not detected; E, estimated value; C, not able to compute RPD due to censored data]

Compound	MRL	Field blank (Pump) (µg/L)	Field blank (Churn) (µg/L)	Equipment blank (ISCO and Churn) (µg/L)	Canal replicate (RPD)	Influent replicate (RPD)	Effluent replicate (RPD)	Influent matrix spike recovery (%)	Effluent matrix spike recovery (%)	Expected matrix spike recovery range (%)
Semivolatile organic compounds (NWQL, Unfiltered, CLLE, GC/MS)										
1,2,4-Trichlorobenzene	0.26	—	—	—	—	—	—	40.6	48.0	34–110
1,2-Dichlorobenzene	0.2	—	—	—	—	—	108	44.7	49.5	36–109
1,2-Diphenylhydrazine	0.3	—	—	—	—	—	—	-21.2	54.5	43–122
1,3-Dichlorobenzene	0.22	—	—	—	—	—	—	44.6	50.5	31–104
1,4-Dichlorobenzene	0.22	—	—	—	—	35.1	43.2	51.0	48.2	31–105
2,4,6-Trichlorophenol	0.34	—	—	—	—	37.8	46.5	48.9	45.1	43–133
2,4-Dichlorophenol	0.36	—	—	—	C	17.5	55.5	-24.2	42.0	39–129
2,4-Dimethylphenol	0.8	—	—	—	—	—	C	53.1	44.1	16–106
2,4-Dinitrophenol	1	—	—	—	—	—	—	-43.8	0.0	34–124
2,4-Dinitrotoluene	0.56	—	—	—	—	—	—	46.8	48.8	43–133
2,6-Dinitrotoluene	0.4	—	—	—	—	—	—	37.7	55.8	49–139
2-Chloronaphthalene	0.16	—	—	—	—	—	—	39.6	51.4	42–120
2-Chlorophenol	0.26	—	—	—	—	—	—	39.1	44.2	38–128
2-Nitrophenol	0.4	—	—	—	—	—	C	46.7	40.8	41–131
4-Bromophenylphenylether	0.24	—	—	—	—	—	—	28.1	52.1	41–119
4-Chloro-3-methylphenol	0.54	—	—	—	—	—	—	61.8	44.5	40–130
4-Chlorophenyl phenyl ether	0.34	—	—	—	—	—	—	28.9	48.3	43–126
4-Nitrophenol	0.52	—	—	—	—	—	—	-15.2	33.4	20–110
Acenaphthene	0.28	—	—	—	—	31.6	48.0	43.3	48.4	41–118
Acenaphthylene	0.3	—	—	—	—	—	—	43.8	48.9	39–121
Benz[a]anthracene	0.26	—	—	—	—	—	—	18.2	42.6	45–123
Benzo[b]fluoranthene	0.3	—	—	—	—	—	—	15.2	34.4	44–124
Benzo[ghi]perylene	0.38	—	—	—	—	—	—	12.4	23.9	31–121
Benzo[k]fluoranthene	0.3	—	—	—	—	—	—	17.2	34.4	41–129
Bis(2-chloroethoxy)methane	0.24	—	—	—	—	—	—	53.9	49.2	48–133
Bis(2-chloroethyl)ether	0.3	—	—	—	—	—	—	50.0	49.7	50–132

Appendix 1. Quality-assurance data for blank, replicate, and matrix spike samples.—Continued

[Recovery values outside expected ranges are in red. MRL, method reporting level; µg/L, micrograms per liter; RPD, Relative Percent Difference; %, percent; NWQL, U.S. Geological Survey National Water Quality Laboratory; CLLE, Continuous liquid-liquid extraction; GC/MS, gas chromatography/mass spectrometry; OGRL, U.S. Geological Survey Organic Geochemistry Research Laboratory; SPE, solid-phase extraction; HPLC/MS, high-performance liquid chromatography/mass spectrometry; ELISA, enzyme-linked immunosorbent assay; —, not detected; E, estimated value; C, not able to compute RPD due to censored data]

Compound	MRL	Field blank (Pump) (µg/L)	Field blank (Churn) (µg/L)	Equipment blank (ISCO and Churn) (µg/L)	Canal replicate (RPD)	Influent replicate (RPD)	Effluent replicate (RPD)	Influent matrix spike recovery (%)	Effluent matrix spike recovery (%)	Expected matrix spike recovery range (%)
Bis(2-chloroisopropyl)ether	0.14	—	—	—	—	—	—	55.1	52.5	47 – 128
Butylbenzyl phthalate	0.9	—	—	—	—	58.6	—	31.9	66.9	62 – 152
Chrysene	0.32	—	—	—	—	—	—	19.3	43.4	43 – 127
Dibenz[a,h]anthracene	0.42	—	—	—	—	—	—	13.9	25.2	25 – 115
Diethyl phthalate	0.62	—	—	—	—	24.0	—	-52.6	51.0	56 – 146
Diethylhexyl phthalate (DEHP)	1.3	—	—	—	C	24.1	—	-50.2	35.1	47 – 137
Dimethyl phthalate	0.36	—	—	—	—	—	—	48.9	43.4	46 – 136
Di-n-butyl phthalate	1	—	—	—	—	33.9	—	30.4	65.6	60 – 150
Di-n-octyl phthalate	0.6	—	—	—	C	2.9	—	11.1	29.7	24 – 114
Fluorene	0.34	—	—	—	—	22.4	53.3	33.7	50.1	46 – 128
Hexachlorobenzene	0.3	—	—	—	—	—	—	20.1	50.1	38 – 117
Hexachlorobutadiene	0.24	—	—	—	—	—	—	27.0	48.9	19 – 88
Hexachlorocyclopentadiene	0.5	—	—	—	—	—	—	11.5	22.1	4 – 87
Hexachloroethane	0.24	—	—	—	—	—	—	23.0	46.0	23 – 95
Indeno[1,2,3-cd]pyrene	0.38	—	—	—	—	—	—	14.0	25.1	30 – 120
Nitrobenzene	0.26	—	—	—	—	—	41.3	49.5	49.1	50 – 134
N-Nitrosodimethylamine (NDMA)	0.24	—	—	—	—	—	—	19.0	38.3	32 – 120
N-Nitrosodi-n-propylamine	0.4	—	—	—	—	—	—	-31.1	61.9	48 – 138
N-Nitrosodiphenylamine	0.28	—	—	—	—	—	—	35.0	52.6	33 – 123
Pentachlorophenol	0.6	—	—	—	—	—	—	129	0.0	40 – 130
Pesticides and pesticide degradates (NWQL, Filtered, SPE, GC/MS)										
1-Naphthol	0.036	—	—	—	—	—	—	71.8	50.4	1 – 43
2,6-Diethylaniline	0.003	—	—	—	—	—	—	110	99.0	48 – 116
2-Chloro-2,6-diethylacetanilide	0.01	—	—	—	—	—	—	136	121	63 – 134
2-Chloro-4-isopropylamino-6-amino-s-triazine (CIAT)	0.006	—	—	—	0.0	—	19.0	81.7	84.4	21 – 89
2-Ethyl-6-methylaniline	0.01	—	—	—	—	—	—	90.1	91.7	48 – 113

Appendix 1. Quality-assurance data for blank, replicate, and matrix spike samples.—Continued

[Recovery values outside expected ranges are in red. MRL, method reporting level; µg/L, micrograms per liter; RPD, Relative Percent Difference; %, percent; NWQL, U.S. Geological Survey National Water Quality Laboratory; CLLE, Continuious liquid-liquid extraction; GC/MS, gas chromatogrraphy/mass spectrometry; OGRL, U.S. Geological Survey Organic Geochemistry Research Laboratory; SPE, solid-phase extraction; HPLC/MS, high-performance liquid chromatogrpahy/mass spectrometry; ELISA, enzyme-linked immunosorbent assay; —, not detected; E, estimated value; C, not able to compute RPD due to censored data]

Compound	MRL	Field blank (Pump) (µg/L)	Field blank (Churn) (µg/L)	Equipment blank (ISCO and Churn) (µg/L)	Canal replicate (RPD)	Influent replicate (RPD)	Effluent replicate (RPD)	Influent matrix spike recovery (%)	Effluent matrix spike recovery (%)	Expected matrix spike recovery range (%)
3,4-Dichloroaniline	0.0042	—	—	—	—	16.3	2.5	22.5	92.6	33 – 102
4-Chloro-2-methylphenol	0.0046	—	—	—	—	—	—	100	95.0	20 – 86
Acetochlor	0.01	—	—	—	—	—	—	123	109	52 – 139
Alachlor	0.008	—	—	—	—	—	—	146	117	58 – 133
Atrazine	0.008	—	—	—	0.0	12.6	1.6	90.5	98.0	69 – 123
Azinphos-methyl	0.12	—	—	—	—	—	—	141	113	18 – 134
Azinphos-methyl-oxon	0.042	—	—	—	—	—	—	36.5	45.2	1 – 121
Benfluralin	0.014	—	—	—	—	—	—	79.6	77.2	32 – 96
Carbaryl	0.06	—	—	—	—	9.8	3.7	142	159	46 – 161
Chlorpyrifos	0.0018	—	—	—	—	11.5	10.0	93.4	88.2	36 – 127
Chlorpyrifos, oxygen analog	0.03	—	—	—	—	—	—	35.7	35.1	1 – 93
cis-Permethrin	0.01	—	—	—	—	—	—	36.0	79.2	18 – 68
Cyfluthrin	0.016	—	—	—	—	—	—	-33.1	106	23 – 66
Cypermethrin	0.02	—	—	—	—	—	—	-2.8	139	19 – 69
Dacthal	0.0076	—	—	—	—	—	—	93.6	97.7	71 – 124
Desulfinylfipronil	0.012	—	—	—	—	10.5	0.0	103	108	61 – 140
Desulfinylfipronil amide	0.029	—	—	—	—	—	—	146	149	48 – 143
Diazinon	0.006	—	—	—	—	—	—	97.9	89.8	56 – 118
Diazinon, oxygen analog	0.005	—	—	—	—	—	—	156	123	38 – 144
Dichlorvos	0.02	—	—	—	—	—	—	64.6	19.0	1 – 85
Dicrotophos	0.08	—	—	—	—	—	—	46.5	35.9	10 – 64
Dieldrin	0.008	—	—	—	—	—	—	74.6	76.6	54 – 126
Dimethoate	0.006	—	—	—	—	—	—	54.6	68.0	10 – 62
Ethion	0.008	—	—	—	—	—	—	113	102	33 – 113
Ethion monoxon	0.021	—	—	—	—	—	—	162	158	40 – 123
Fenamiphos	0.03	—	—	—	—	—	—	174	121	36 – 123
Fenamiphos sulfone	0.054	—	—	—	—	—	—	142	121	37 – 111

Appendix 1. Quality-assurance data for blank, replicate, and matrix spike samples.—Continued

[Recovery values outside expected ranges are in red. MRL, method reporting level; µg/L, micrograms per liter; RPD, Relative Percent Difference; %, percent; NWQL, U.S. Geological Survey National Water Quality Laboratory; CLLE, Continuous liquid-liquid extraction; GC/MS, gas chromatogrpahy/mass spectrometry; OGRL, U.S. Geological Survey Organic Geochemistry Research Laboratory; SPE, solid-phase extraction; HPLC/MS, high-performance liquid chromatogrphy/mass spectrometry; ELISA, enzyme-linked immunosorbent assay; —, not detected; E, estimated value; C, not able to compute RPD due to censored data]

Compound	MRL	Field blank (Pump) (µg/L)	Field blank (Churn) (µg/L)	Equipment blank (ISCO and Churn) (µg/L)	Canal replicate (RPD)	Influent replicate (RPD)	Effluent replicate (RPD)	Influent matrix spike recovery (%)	Effluent matrix spike recovery (%)	Expected matrix spike recovery range (%)
Fenamiphos sulfoxide	0.08	—	—	—	—	—	—	41.1	19.3	1 – 58
Fipronil	0.018	—	—	—	—	8.2	0.6	148	130	24 – 147
Fipronil sulfide	0.012	—	—	—	—	9.2	7.3	111	105	33 – 146
Fipronil sulfone	0.024	—	—	—	—	—	—	102	97.7	33 – 116
Fonofos	0.0048	—	—	—	—	—	—	101	93.1	53 – 116
Hexazinone	0.008	—	—	—	8.5	—	—	101	84.5	23 – 99
Iprodione	0.014	—	—	—	—	—	—	113	102	9 – 112
Isofenphos	0.006	—	—	—	—	—	—	161	119	40 – 140
Malaoxon	0.022	—	—	—	—	—	—	135	102	44 – 137
Malathion	0.016	—	—	—	—	—	—	142	127	50 – 135
Metalaxyl	0.014	—	—	—	—	—	—	78.1	99.7	54 – 135
Methidathion	0.012	—	—	—	—	—	—	99.3	112	54 – 125
Metolachlor	0.02	—	—	—	—	—	—	118	111	55 – 132
Metribuzin	0.012	—	—	—	—	—	—	126	119	42 – 118
Myclobutanil	0.01	—	—	—	—	—	—	115	112	45 – 129
Paraoxon-methyl	0.014	—	—	—	—	—	—	117	66.0	14 – 123
Parathion-methyl	0.008	—	—	—	—	—	—	127	106	33 – 121
Pendimethalin	0.012	—	—	—	—	—	—	125	116	47 – 127
Phorate	0.02	—	—	—	—	—	—	85.2	59.8	21 – 104
Phorate oxygen analog	0.027	—	—	—	—	—	—	196	145	43 – 157
Phosmet	0.07	—	—	—	—	—	—	92.9	58.9	1 – 49
Phosmet oxon	0.0079	—	—	—	—	—	—	28.4	26.9	1 – 66
Prometon	0.012	—	—	—	—	—	—	0.0	110	49 – 130
Prometryn	0.006	—	—	—	—	—	—	117	114	50 – 140
Propyzamide	0.0036	—	—	—	—	—	—	111	102	46 – 132
Simazine	0.006	—	—	—	—	—	—	103	113	52 – 131
Tebuthiuron	0.028	—	—	—	—	—	—	-3.5	166	39 – 198

Appendix 1. Quality-assurance data for blank, replicate, and matrix spike samples.—Continued

[Recovery values outside expected ranges are in red. MRL, method reporting level; µg/L, micrograms per liter; RPD, Relative Percent Difference; %, percent; NWQL, U.S. Geological Survey National Water Quality Laboratory; CLLE, Continuous liquid-liquid extraction; GC/MS, gas chromatogrpahy/mass spectrometry; OGRL, U.S. Geological Survey Organic Geochemistry Research Laboratory; SPE, solid-phase extraction; HPLC/MS, high-performance liquid chromatogrpahy/mass spectrometry; ELISA, enzyme-linked immunosorbent assay; —, not detected; E, estimated value; C, not able to compute RPD due to censored data]

Compound	MRL	Field blank (Pump) (µg/L)	Field blank (Churn) (µg/L)	Equipment blank (ISCO and Churn) (µg/L)	Canal replicate (RPD)	Influent replicate (RPD)	Effluent replicate (RPD)	Influent matrix spike recovery (%)	Effluent matrix spike recovery (%)	Expected matrix spike recovery range (%)
Terbufos	0.018	—	—	—	—	—	—	103	97.0	39 – 110
Terbufos oxygen analog sulfone	0.045	—	—	—	—	—	—	172	213	37 – 146
Terbuthylazine	0.006	—	—	—	—	0.8	4.9	85.2	96.2	67 – 129
Tribufos	0.018	—	—	—	—	—	—	1.4	73.9	19 – 87
Wastewater-indicator compounds (NWQL, Unfiltered, CLLE, GC/MS)										
1-Methylnaphthalene	0.02	—	—	—	—	59.3	—	86.7	86.1	49 – 113
2,2',4,4'-Tetrabromodiphenylether (PBDE 47)	0.02	—	—	—	—	—	—	25.4	61.3	33 – 123
2,6-Dimethylnaphthalene	0.02	—	—	—	0.0	31.5	—	60.3	85.1	50 – 110
2-Methylnaphthalene	0.02	—	—	—	—	52.8	—	114	85.7	48 – 113
3,4-Dichlorophenyl isocyanate	0.16	—	—	—	—	—	61.2	0.0	29.8	97 – 187
3-*beta*-Coprostanol	0.38	—	—	—	—	7.6	—	371	64.6	34 – 124
3-Methyl-1H-indole (Skatole)	0.02	—	—	—	—	73.9	28.6	132	89.8	48 – 130
3-*tert*-Butyl-4-hydroxyanisole (BHA)	0.08	—	—	—	—	39.5	—	123	122	8 – 98
4-Cumylphenol	0.02	—	—	—	—	26.3	—	80.1	103	50 – 130
4-n-Octylphenol	0.01	—	—	—	—	—	—	79.8	112	46 – 128
4-Nonylphenol (total, NP)	1.2	—	—	—	—	29.0	16.8	93.1	107	50 – 140
4-Nonylphenol diethoxylate (NP$_2$EO)	0.8	—	—	—	—	11.3	—	32.9	81.0	47 – 137
4-Nonylphenol monoethoxylate (NP$_1$EO)	1.3	E0.30	—	E0.11	C	C	—	139	104	52 – 142
4-Octylphenol diethoxylate (OP$_2$EO)	0.1	—	—	—	—	—	52.7	-285	116	43 – 133
4-Octylphenol monoethoxylate (OP$_1$EO)	0.3	—	—	—	—	35.7	C	113	116	50 – 140
4-*tert*-Octylphenol (OP)	0.11	—	—	—	—	32.6	19.4	236	116	52 – 142
5-Methyl-1H-benzotriazole	0.16	—	—	—	—	—	—	117	82.7	33 – 123
Acetophenone	0.07	—	—	—	—	37.2	26.0	110	110	66 – 142
Anthracene	0.01	—	—	—	—	28.3	—	58.6	92.0	54 – 118
Anthraquinone	0.02	—	—	—	—	—	43.2	114	107	52 – 128
Benzo[a]pyrene	0.01	—	—	—	—	—	—	28.0	62.6	49 – 116

Appendix 1. Quality-assurance data for blank, replicate, and matrix spike samples.—Continued

[Recovery values outside expected ranges are in red. MRL, method reporting level; µg/L, micrograms per liter; RPD, Relative Percent Difference; %, percent; NWQL, U.S. Geological Survey National Water Quality Laboratory; CLLE, Continiuous liquid-liquid extraction; GC/MS, gas chromatogrphay/mass spectrometry; OGRL, U.S. Geological Survey Organic Geochemistry Research Laboratory; SPE, solid-phase extraction; HPLC/MS, high-performance liquid chromatogrphay/mass spectrometry; ELISA, enzyme-linked immunosorbent assay; —, not detected; E, estimated value; C, not able to compute RPD due to censored data]

Compound	MRL	Field blank (Pump) (µg/L)	Field blank (Churn) (µg/L)	Equipment blank (ISCO and Churn) (µg/L)	Canal replicate (RPD)	Influent replicate (RPD)	Effluent replicate (RPD)	Influent matrix spike recovery (%)	Effluent matrix spike recovery (%)	Expected matrix spike recovery range (%)
Benzophenone	0.08	—	E0.06	—	—	26.7	26.7	102	108	54 – 143
beta–Sitosterol	0.8	—	—	—	—	8.7	—	36.9	27.4	32 – 122
beta–Stigmastanol	0.8	—	—	—	22.2	7.0	—	28.7	22.2	31 – 121
Bisphenol A	0.02	—	—	—	—	—	—	110	113	43 – 133
Bromacil	0.08	—	—	—	—	—	—	78.1	98.7	51 – 133
Bromoform	0.08	—	—	—	—	53.0	16.6	56.1	60.8	44 – 111
Camphor	0.04	—	—	—	—	33.1	—	131	93.8	55 – 124
Carbazole	0.01	—	—	—	—	44.3	—	108	89.5	51 – 127
Cholesterol	0.3	—	—	—	20.3	15.3	—	469	48.3	33 – 123
Cotinine	0.04	—	—	—	—	20.8	—	80.7	75.6	12 – 102
Fluoranthene	0.01	—	—	—	—	35.0	C	54.6	91.6	53 – 125
Galaxolide (HHCB)	0.04	—	E0.025	—	—	19.1	10.4	218	148	53 – 120
Indole	0.02	—	—	—	—	48.7	—	17.8	55.0	45 – 135
Isoborneol	0.04	—	—	—	—	40.9	—	109	102	41 – 131
Isophorone	0.02	—	—	—	—	—	28.9	105	99.9	55 – 125
Isopropylbenzene	0.02	—	—	—	—	42.1	—	42.6	56.1	14 – 104
Isoquinoline	0.02	—	—	—	—	—	—	97.1	94.7	40 – 130
Limonene	0.2	—	—	—	—	63.5	—	72.7	50.8	7 – 93
Menthol	0.16	—	—	—	—	46.7	4.6	172	89.7	45 – 135
Methyl salicylate	0.04	—	—	—	—	41.3	58.0	106	93.5	47 – 129
N,N-diethyl-meta-toluamide (DEET)	0.02	—	E0.003	—	—	42.2	37.9	212	110	61 – 151
Naphthalene	0.01	—	—	—	—	48.5	4.4	131	81.8	52 – 118
para-Cresol	0.04	—	—	—	—	55.2	—	94.4	109	54 – 130
Phenanthrene	0.01	—	—	—	—	40.4	—	118	89.0	53 – 118
Phenol	0.08	—	0.51	—	—	42.3	167	54.4	135	45 – 116
Pyrene	0.01	—	—	—	—	9.0	—	13.6	92.8	53 – 126
Tetrachloroethylene	0.08	—	—	—	—	25.9	—	25.4	36.0	2 – 77

Appendix 1. Quality-assurance data for blank, replicate, and matrix spike samples.—Continued

[Recovery values outside expected ranges are in red. MRL, method reporting level; µg/L, micrograms per liter; RPD, Relative Percent Difference; %, percent; NWQL, U.S. Geological Survey National Water Quality Laboratory; CLLE, Continuous liquid-liquid extraction; GC/MS, gas chromatogrpahy/mass spectrometry; OGRL, U.S. Geological Survey Organic Geochemistry Research Laboratory; SPE, solid-phase extraction; HPLC/MS, high-performance liquid chromatogrpahy/mass spectrometry; ELISA, enzyme-linked immunosorbent assay; —, not detected; E, estimated value; C, not able to compute RPD due to censored data]

Compound	MRL	Field blank (Pump) (µg/L)	Field blank (Churn) (µg/L)	Equipment blank (ISCO and Churn) (µg/L)	Canal replicate (RPD)	Influent replicate (RPD)	Effluent replicate (RPD)	Influent matrix spike recovery (%)	Effluent matrix spike recovery (%)	Expected matrix spike recovery range (%)
Tonalide (AHTN)	0.02	—	—	—	—	61.3	13.6	101	88.3	51–125
Tri(2-butoxyethyl)phosphate (TBEP)	0.32	—	—	—	—	10.7	25.0	98.3	94.1	47–137
Tri(2-chloroethyl)phosphate (TCEP)	0.08	—	—	E0.015	—	45.9	24.0	96.7	97.3	49–135
Tri(dichlorisopropyl)phosphate (TDIP)	0.16	—	—	—	—	17.9	32.6	79.8	102	49–134
Tributyl phosphate (TBP)	0.02	—	—	—	—	37.4	20.3	83.5	126	56–146
Triclosan	0.16	E0.009	—	—	—	32.4	—	79.3	93.2	48–127
Triethyl citrate (Ethyl citrate)	0.02	—	E0.011	—	—	43.6	29.0	126	111	46–136
Triphenyl phosphate	0.04	—	—	E0.01	—	12.8	27.3	47.0	89.0	52–129
Pharmaceutical compounds (NWQL, Filtered, SPE, HPLC/MS)										
1,7-Dimethylxanthine (Wastewater Indicator Compound)	0.1	—	—	—	—	1.3	—	-508	31.7	23–151
Acetaminophen	0.12	—	—	—	1.2	2.0	C	-400	8.2	10–152
Albuterol	0.08	—	—	—	—	—	—	96.4	27.6	25–148
Caffeine (Wastewater Indicator Compound)	0.06	—	—	—	15.0	1.3	106	12.2	50.4	52–147
Codeine	0.046	—	—	—	—	C	38.2	35.3	53.3	36–129
Dehydronifedipine	0.08	—	—	—	—	—	17.3	46.5	82.1	39–151
Diltiazem	0.02	—	—	—	—	—	—	0.0	18.2	7–86
Diphenhydramine	0.058	—	—	—	—	—	14.8	0.0	13.5	35–100
Thiabendazole	0.06	—	—	—	—	C	—	-1.7	9.1	11–141
Warfarin	0.08	—	—	—	—	—	—	22.8	39.4	0–129
Antibiotics (OGRL, Filtered, SPE, LC/MS)										
Azithromycin	0.005	—	—	—	—	C	11.6	-1	82	
Carbamazepine (Pharmaceutical)	0.005	—	—	—	—	2.6	13.6	136	110	
Chloramphenicol	0.1	—	—	—	—	—	—	0.0	0.0	
Chlortetracycline	0.01	—	—	—	—	—	—	—	—	
Ciproflaxacin	0.005	—	—	—	—	60.2	53.8	432	94	
Doxycycline	0.01	—	—	—	—	—	—	121	117	

Appendix 1. Quality-assurance data for blank, replicate, and matrix spike samples.—Continued

[Recovery values outside expected ranges are in red. MRL, method reporting level; µg/L, micrograms per liter; RPD, Relative Percent Difference; %, percent; NWQL, U.S. Geological Survey National Water Quality Laboratory; CLLE, Continuous liquid-liquid extraction; GC/MS, gas chromatogrpahy/mass spectrometry; OGRL, U.S. Geological Survey Organic Geochemistry Research Laboratory; SPE, solid-phase extraction; HPLC/MS, high-performance liquid chromatogrpahy/mass spectrometry; ELISA, enzyme-linked immunosorbent assay; —, not detected; E, estimated value; C, not able to compute RPD due to censored data]

Compound	MRL	Field blank (Pump) (µg/L)	Field blank (Churn) (µg/L)	Equipment blank (ISCO and Churn) (µg/L)	Canal replicate (RPD)	Influent replicate (RPD)	Effluent replicate (RPD)	Influent matrix spike recovery (%)	Effluent matrix spike recovery (%)	Expected matrix spike recovery range (%)
Enrofloxacin	0.005	—	—	—	—	—	—	278	114	—
epi-Chlortetracycline	0.01	—	—	—	—	—	—	—	—	—
epi-iso-Chlortetracycline	0.01	—	—	—	—	—	—	—	—	—
epi-Oxytetracycline	0.01	—	—	—	—	—	—	—	—	—
epi-Tetracycline	0.01	—	—	—	—	—	—	—	—	—
Erythromycin	0.008	—	—	—	—	C	16.2	64.5	136	—
Erythromycin-H$_2$O	0.008	—	—	—	—	83.9	10.0	30.5	106	—
Ibuprofen (Pharmaceutical)	0.005	—	—	—	—	—	—	0.0	0.0	—
iso-Chlortetracycline	0.01	—	—	—	—	—	—	—	—	—
Lincomycin	0.005	—	—	—	—	—	—	212	172	—
Lomefloxacin	0.005	—	—	—	—	—	—	180	115	—
Norfloxacin	0.005	—	—	—	—	—	—	304	135	—
Ofloxacin	0.005	—	—	—	—	50.9	68.7	757	98	—
Ormetoprim	0.005	—	—	—	—	—	—	65.5	89.5	—
Oxytetracycline	0.01	—	—	—	—	—	—	138	97.5	—
Roxithromycin	0.005	—	—	—	—	—	—	117	104	—
Sarafloxacin	0.005	—	—	—	—	—	—	216	86	—
Sulfachloropyridazine	0.005	—	—	—	—	—	—	107	101	—
Sulfadiazine	0.1	—	—	—	—	21.3	16.7	78	65.5	—
Sulfadimethoxine	0.005	—	—	—	—	—	—	141	193	—
Sulfamethazine	0.005	—	—	—	—	—	—	91.5	117	—
Sulfamethoxazole	0.005	—	—	—	—	53.9	5.6	59	82.5	—
Sulfathiazole	0.05	—	—	—	—	—	—	105	92.5	—
Tetracycline	0.01	—	—	—	—	22.2	—	123	114	—
Trimethoprim	0.005	—	—	—	—	140	14.3	-75.5	88	—
Tylosin	0.01	—	—	—	—	—	—	211	99	—
Virginiamycin	0.005	—	—	—	—	—	—	0.0	0.0	—

Appendix 1.   Quality-assurance data for blank, replicate, and matrix spike samples.—Continued

[Recovery values outside expected ranges are in red. MRL, method reporting level; µg/L, micrograms per liter; RPD, Relative Percent Difference; %, percent; NWQL, U.S. Geological Survey National Water Quality Laboratory; CLLE, Continuous liquid-liquid extraction; GC/MS, gas chromatogrpahy/mass spectrometry; OGRL, U.S. Geological Survey Organic Geochemistry Research Laboratory; SPE, solid-phase extraction; HPLC/MS, high-performance liquid chromatogrpahy/mass spectrometry; ELISA, enzyme-linked immunosorbent assay; —, not detected; E, estimated value; C, not able to compute RPD due to censored data]

Compound	MRL	Field blank (Pump) (µg/L)	Field blank (Churn) (µg/L)	Equipment blank (ISCO and Churn) (µg/L)	Canal replicate (RPD)	Influent replicate (RPD)	Effluent replicate (RPD)	Influent matrix spike recovery (%)	Effluent matrix spike recovery (%)	Expected matrix spike recovery range (%)
Hormone (OGRL, Filtered, ELISA)										
17-*beta*-estradiol (EZ)	—			—	—	28.6	8.7	120	88.0	
Median					4.9	31.5	20.3	79.7	88.3	
Minimum					0.0	0.8	0.0	-508	0.0	
Maximum					22.2	140	167	757	213	
Median (SVOC)								31.1	48.1	
Median (pesticides and pesticide degradates)								101.4	101.6	
Median (WWIC)								95.5	93.0	
Median (pharmaceutical compounds)								6.1	29.7	
Median (antibiotics)								116.5	99.0	

Appendix 2. Percent recovery values for surrogate compounds added to groundwater, surface-water, effluent, and influent samples at the U.S. Geological Survey National Water Quality Laboratory.

[Values in percent recovery; yyyymmdd, year, month, day; shading indicates different laboratory methods; Abbreviations: R, Replicate; GW, Groundwater; SW, Surface water; WE, Effluent Wastewater; WI, Influent Wastewater; HS, Homestead; Eff, effluent; —, not applicable; RSD, relative standard deviation]

Site identifier	Matrix type	Sample date (yyyymmdd)	Sample time	2,4,6-Tribromophenol	2-Fluorobiphenyl	2-Fluorophenol	Nitrobenzene-d_5	Phenol-d_5	Terphenyl-d_{14}	alpha-HCH-d_6	Diazinon-d_{10}	Carbamazepine-d_{10}	Ethyl nicotinate-d_4	Bisphenol A-d_3	Caffeine-C^{13}	Decafluorobiphenyl	Fluoranthene-d_{10}
Expected range:				38-120	46-125	33-120	54-130	26-120	46-120	70-116	40-156	61-147	63-127	25-120	50-120	35-120	60-125
MW-3	GW	20080811	0900	96.1	88.1	82.0	90.9	63.6	89.1	138	91.6	71.4	80.0	118	61.4	54.3	70.7
MW-3	GW	20090210	1300	114	87.7	89.7	92.6	80.2	91.5	155	99.6	41.1	66.6	59.8	55.1	43.6	61.1
MW-2	GW	20080811	1100	94.1	89.7	74.7	96.8	55.1	92.1	143	94.0	80.8	80.1	105	62.7	59.0	70.5
MW-2	GW	20090211	1500	89.4	81.5	49.8	96.1	44.3	78.8	175	101	34.1	66.5	67.9	54.7	47.8	69.5
MW-1	GW	20080811	1500	94.1	85.4	69.8	94.4	58.2	84.3	128	93.6	87.2	89.8	111	64.5	56.1	71.5
MW-1	GW	20090210	1530	117	88.6	91.0	96.5	79.2	86.5	170	102	43.4	69.2	28.4	29.0	24.2	32.6
MC-1	SW	20080812	1200	75.6	74.4	53.9	78.0	38.2	72.6	93.8	99.6	116	87.3	120	63.0	49.3	70.4
MC-1	SW	20091209	1130	89.7	76.6	58.7	89.6	47.1	48.5	95.4	94.4	56.8	75.7	75.4	59.6	45.6	53.5
MC-2	SW	20080812	1600	83.3	82.2	64.8	85.9	49.2	75.9	97.5	91.3	119	88.5	134	76.3	55.3	78.3
MC-2	SW	20091209	1230	90.5	69.8	57.7	82.7	46.3	45.4	96.4	100	59.4	78.4	48.8	40.9	32.7	37.8
SC	SW	20080709	0900	70.1	53.3	52.0	63.9	39.2	51.1	100	105	95.4	97.2	71.1	78.0	49.0	77.9
SC-R	SW	20080709	0905	95.2	80.8	70.3	93.7	55.1	84.0	97.1	110	93.7	95.1	67.8	75.2	45.0	80.4
HS-Eff	WE	20090210	0200	130	91.9	83.3	103.5	66.0	74.1	108	184	27.5	65.5	37.4	34.9	27.6	43.7
HS-Eff	WE	20091020	1000	113	81.4	81.1	86.9	68.9	65.0	83.6	142	43.5	55.7	48.0	58.3	39.9	58.1
HS-Eff	WE	20091020	1001	—	—	—	—	—	—	—	—	—	—	85.7	86.7	51.7	82.0
HS-Eff-R	WE	20091020	1005	62.3	56.5	42.3	60.2	42.8	50.7	79.5	136	52.2	78.3	71.2	76.8	36.6	75.9
HS-Eff	WE	20091020	1011	—	—	—	—	—	—	—	—	40.3	66.4	—	—	—	—
HS-Eff	WE	20091020	1021	47.5	48.7	37.4	50.3	35.7	41.6	—	—	—	—	—	—	—	—
HS-Eff	WE	20091020	1031	—	—	—	—	—	—	81.8	131	—	—	—	—	—	—
SD1-Eff	WE	20090128	0400	108	74.3	72.5	88.3	64.6	42.1	99.7	159	14.8	75.4	67.4	72.2	67.0	63.6
SD1-Eff	WE	20091013	1400	125	89.7	75.6	110	75.0	66.4	87.9	175	20.5	61.2	104	71.7	41.6	67.3
SD2-Eff	WE	20090128	0400	33.6	27.0	14.8	28.6	15.1	16.8	104	162	13.9	69.1	79.3	88.5	77.7	69.6
SD2-Eff	WE	20091013	1400	117	82.7	80.7	102	75.0	57.1	87.8	169	16.0	24.3	119	93.7	53.3	85.5
CD1-Eff	WE	20090420	1000	123	93.5	94.4	103	74.7	44.3	88.9	161	9.3	60.0	—	—	—	—
CD1-Eff	WE	20090929	1000	37.5	66.1	26.9	77.0	26.7	31.6	80.1	127	17.3	52.8	54.3	57.8	49.2	61.5
CD2-Eff	WE	20090427	1600	108	77.3	73.1	94.9	60.4	38.6	93.8	170	8.4	49.8	91.0	63.4	43.9	58.1
CD2-Eff	WE	20090929	1000	88.5	75.2	54.1	90.8	47.3	39.4	86.4	133	13.7	47.0	30.0	45.9	33.4	42.8
ND-Eff	WE	20081209	1400	99.7	73.0	69.0	90.7	52.2	37.5	90.4	141	18.8	70.1	64.2	61.8	51.1	56.3

Appendix 2. Percent recovery values for surrogate compounds added to groundwater, surface-water, effluent, and influent samples at the U.S. Geological Survey National Water Quality Laboratory.—Continued

[Values in percent recovery; yyyymmdd, year, month, day; shading indicates different laboratory methods; Abbreviations: R, Replicate; GW, Groundwater; SW, Surface water; WE, Effluent Wastewater; WI, Influent Wastewater; HS, Homestead; Eff, effluent; —, not applicable; RSD, relative standard deviation]

Site identifier	Matrix type	Sample date (yyyymmdd)	Sample time	2,4,6-Tribromophenol	2-Fluorobiphenyl	2-Fluorophenol	Nitrobenzene-d_5	Phenol-d_5	Terphenyl-d_{14}	alpha-HCH-d_6	Diazinon-d_{10}	Carbamazepine-d_{10}	Ethyl nicotinate-d_4	Bisphenol A-d_3	Caffeine-C^{13}	Decafluorobiphenyl	Fluoranthene-d_{10}
Expected range:				38-120	46-125	33-120	54-130	26-120	46-120	70-116	40-156	61-147	63-127	25-120	50-120	35-120	60-125
ND-Eff	WE	20090922	1600	90.4	85.1	49.6	102	39.9	51.6	74.8	119	12.7	49.0	80.4	63.5	52.1	59.1
HS-Inf	WI	20090210	0000	136	72.1	101	100	110	52.9	94.2	137	13.8	61.9	54.9	24.8	14.6	10.9
HS-Inf	WI	20091020	0800	70.3	35.9	36.9	60.2	41.0	24.8	80.3	145	19.1	41.2	0.0	50.5	19.3	24.7
HS-Inf	WI	20091020	0801	—	—	—	—	—	—	—	—	—	—	0.0	82.3	30.9	37.9
HS-Inf-R	WI	20091020	0805	100	50.0	64.7	80.0	66.1	40.1	77.9	137	22.8	46.9	0.0	80.6	28.7	36.5
HS-Inf	WI	20091020	0811	—	—	—	—	—	—	—	—	17.6	41.7	—	—	—	—
HS-Inf	WI	20091020	0821	59.7	34.7	22.2	51.5	23.0	18.0	—	—	—	—	—	—	—	—
HS-Inf	WI	20091020	0831	—	—	—	—	—	—	80.9	137	—	—	—	—	—	—
SD1-Inf	WI	20090128	0000	87.1	53.6	40.8	87.0	58.5	27.1	105	160	22.1	76.4	0.0	0.0	47.0	32.1
SD1-Inf	WI	20091013	1200	21.1	22.0	15.8	32.8	17.3	23.6	84.2	144	10.5	39.9	0.0	59.5	21.4	29.2
SD2-Inf	WI	20090128	0000	107	59.5	50.0	83.7	73.4	33.7	91.4	156	16.6	63.5	0.0	0.0	40.4	71.8
SD2-Inf	WI	20091013	1200	78.9	50.6	50.1	93.0	48.0	44.3	85.2	137	11.7	38.2	0.0	76.2	24.0	37.0
CD1-Inf	WI	20090420	0800	52.2	81.9	55.3	98.9	33.8	—	91.1	158	8.0	49.5	—	65.9	—	—
CD1-Inf	WI	20090929	0800	68.0	51.2	46.1	84.2	46.6	36.8	82.1	148	22.9	48.1	0.0	65.9	35.0	37.8
CD2-Inf	WI	20090427	1400	92.8	64.3	78.9	124	62.7	50.3	96.2	151	7.2	32.5	0.0	32.0	20.7	27.1
CD2-Inf	WI	20090929	0800	72.2	50.5	51.1	90.6	44.7	30.9	82.9	151	19.6	42.3	0.0	64.2	37.5	42.0
ND-Inf	WI	20081209	1000	98.6	48.2	64.8	98.6	67.3	30.8	80.1	111	13.0	71.6	0.0	97.4	38.9	35.0
ND-Inf	WI	20090922	1400	82.0	51.4	59.5	94.0	55.6	26.1	74.2	132	10.2	35.2	0.0	96.3	40.2	41.3

www.ingramcontent.com/pod-product-compliance
Lightning Source LLC
Chambersburg PA
CBHW081849170526
45167CB00007B/2939